Two-Dimensional Random Walk

The main subject of this introductory book is simple random walk on the integer lattice, with special attention to the two-dimensional case. This fascinating mathematical object is the point of departure for an intuitive and richly illustrated tour of related topics at the active edge of research. The book starts with three different proofs of the recurrence of the two-dimensional walk, via direct combinatorial arguments, electrical networks, and Lyapunov functions. Then, after reviewing some relevant potential-theoretic tools, the reader is guided towards the relatively new topic of random interlacements – which can be viewed as a "canonical soup" of nearest-neighbour loops through infinity – again with emphasis on two dimensions. On the way, readers will visit conditioned simple random walks – which are the "noodles" in the soup – and also discover how Poisson processes of infinite objects are constructed and review the recently introduced method of soft local times. Each chapter ends with many exercises, making the book suitable for courses and for independent study.

Serguei Popov works on questions related to random walks (also in random environments), random interlacements and others. He also wrote (together with Mikhail Menshikov and Andrew Wade) a book on the Lyapunov functions method for Markov chains. Recent works of the author on random interlacements (including the two-dimensional case) attracted considerable interest in the probabilistic community. Perhaps his most important recent contribution is the soft local times method for constructing couplings of stochastic processes, developed in a joint work with Augusto Teixeira. This method not only permitted strong advances in the field of random interlacements but also proved its usefulness in other topics.

INSTITUTE OF MATHEMATICAL STATISTICS
TEXTBOOKS

IMS Textbooks give introductory accounts of topics of current concern suitable for advanced courses at master's level, for doctoral students and for individual study. They are typically shorter than a fully developed textbook, often arising from material created for a topical course. Lengths of 100–290 pages are envisaged. The books typically contain exercises.

In collaboration with the International Society for Bayesian Analysis (ISBA), selected volumes in the IMS Textbooks series carry the "with ISBA" designation at the recommendation of the ISBA editorial representative.

Two-Dimensional Random Walk

From Path Counting to Random Interlacements

SERGUEI POPOV

University of Porto

CAMBRIDGE
UNIVERSITY PRESS

CAMBRIDGE
UNIVERSITY PRESS

University Printing House, Cambridge CB2 8BS, United Kingdom

One Liberty Plaza, 20th Floor, New York, NY 10006, USA

477 Williamstown Road, Port Melbourne, VIC 3207, Australia

314-321, 3rd Floor, Plot 3, Splendor Forum, Jasola District Centre, New Delhi – 110025, India

79 Anson Road, #06-04/06, Singapore 079906

Cambridge University Press is part of the University of Cambridge.

It furthers the University's mission by disseminating knowledge in the pursuit of education, learning, and research at the highest international levels of excellence.

www.cambridge.org
Information on this title: www.cambridge.org/9781108472456
DOI: 10.1017/9781108680134

First published 2021

A catalogue record for this publication is available from the British Library.

Library of Congress Cataloging-in-Publication Data
Names: Popov, Serguei, 1972– author.
Title: Two-dimensional random walk : from path counting to random interlacements / Serguei Popov, University of Porto.
Description: Cambridge, United Kingdom ; New York, NY : Cambridge University Press, 2020. |
Series: Institute of mathematical statistics textbooks |
Includes bibliographical references and index.
Identifiers: LCCN 2020020947 (print) | LCCN 2020020948 (ebook) |
ISBN 9781108472456 (hardback) | ISBN 9781108459693 (paperback) |
ISBN 9781108680134 (epub)
Subjects: LCSH: Random walks (Mathematics)
Classification: LCC QA274.73 .P66 2020 (print) | LCC QA274.73 (ebook) |
DDC 519.2/82–dc23
LC record available at https://lccn.loc.gov/2020020947
LC ebook record available at https://lccn.loc.gov/2020020948

ISBN 978-1-108-47245-6 Hardback
ISBN 978-1-108-45969-3 Paperback

Contents

Preface

What does it look like when a mathematician explains something to a fellow mathematician? Everyone knows: lots of writing on the blackboard, lots of intuition flying around, and so on. It is not surprising that mathematicians often prefer a conversation with a colleague to "simply" reading a book. So, in view of this, my initial goal was to write a book as if I were just *explaining* things to a colleague or a research student. In such a book, there should be a lot of pictures and plenty of detailed explanations, so that the reader would hardly have any questions left. After all, wouldn't it be nice if a person (hmm... well, a mathematician) could just read it in a bus (bed, park, sofa, etc.) and still learn some ideas from contemporary mathematics? I have to confess that, unfortunately, as attested by many early readers, I have not always been successful in creating a text with the aforementioned properties. Still, I hope that at least some pieces will be up to the mission.

To explain my motivation, consider the following well-known fact: frequently, the proof of a mathematical result is difficult, containing lots of technicalities which are hard to follow. It is not uncommon that people struggle to understand such proofs without first getting a "general idea" about what is going on. Also, one forgets technicalities[1] but general ideas remain (and if the ideas are retained, the technical details can usually be reconstructed with some work). So, in this book the following approach is used. I will *always* prefer to explain the intuition first. If the proof is instructive and not too long, it will be included. Otherwise, I will let the interested reader look up the details in other books and/or papers.

The approach can be characterized as striving to understand all things through direct probabilistic intuition. Yes, I am aware that this is not always possible. Nonetheless, when facing a complex task, it is frequently easier to

---

[1] Even of one's own proofs, as the author has learned on quite a few occasions.

tackle it using familiar tools[2] (even in a non-optimal way) as much as possible than to employ other (possibly more adequate) tools one is unfamiliar with. Also, advanced approaches applied to basic tasks have a tendency to "hide" what is *really* going on (one becomes enchanted with the magic, while still not being able to perform it oneself).

This book revolves around two-dimensional simple random walk, which is not actually so simple, but in fact a rich and fascinating mathematical object. Our purpose here is not to provide a complete treatment of that object, but rather to make an interesting tour around it. In the end, we will come to the relatively new topic of random interlacements (which can be viewed as "canonical" nearest-neighbour loops through infinity). Also, on the way there will be several "side-quests": we will take our time to digress to some related topics which are somewhat underrepresented in the literature, such as Lyapunov functions and Doob's *h*-transforms for Markov chains.

Intended audience

I expect this book to be of interest to research students and postdocs working with random walks, and to mathematicians in neighbouring fields. Given the approach I take, it is better suited to those who want to "get the intuition first", i.e. first obtain a general idea of what is going on, and only after that pass to technicalities. I am aware that not everybody likes this approach, but I hope that the book will find its audience. Although this book is designed primarily for self-study, it can also be used for a one-semester course on additional topics in Markov chains.

The technical prerequisites are rather mild. The material in the book will be at a level accessible to readers familiar with the basic concepts of probability theory, including convergence of random variables and uniform integrability, with also some background in martingales and Markov chains (at the level of [44], for example). The book is meant to be mostly self-contained (and we recall all necessary definitions and results in Chapter 1).

Many topics in the book are treated at length in the literature, e.g. [41, 63, 71, 91]; on the other hand, we also discuss some recent advances (namely, soft local times and two-dimensional random interlacements) that have not been covered in other books. In any case, the main distinguishing feature of this book is not its content, but rather the way it is presented.

---

[2] In Russia, the ability to build a log house using *only* an axe was considered proof of a carpenter's craftsmanship.

Overview of content

The content of the book is described here. Each chapter (except for the introduction) ends with a list of exercises, and a section with hints and solutions to selected exercises appears at the end of the book. A note about the exercises: they are mostly *not* meant to be easily solved during a walk in the park; the purpose of at least some of them is to guide an interested reader who wants to dive deeper into the subject.

1. *Basic definitions.* We recall here some basic definitions and facts for Markov chains and martingales, mainly for reference purposes.
2. *Recurrence of two-dimensional simple random walk.* First, we recall two well-known proofs of recurrence of simple random walk in two dimensions: the classical combinatorial proof and the proof with electrical networks. We then observe that the first proof relies heavily on specific combinatorics and so is very sensitive to small changes in the model's parameters, and the second proof applies only to reversible Markov chains. Then, we present a very short introduction to the Lyapunov function method – which neither requires reversibility nor is sensitive to small perturbations of transition probabilities.
3. *Some potential theory for simple random walks.* This chapter gives a gentle introduction to the potential theory for simple random walks, first in the transient case ($d \geq 3$), and then in two dimensions. The idea is to recall and discuss the basic concepts (such as Green's function, potential kernel, harmonic measure) needed in the rest of the book; this chapter is not intended to provide a profound treatment of the subject.
4. *Simple random walk conditioned on not hitting the origin.* Here, we first recall the idea of Doob's *h*-transform, which permits us to represent a conditioned (on an event of not hitting some set) Markov chain as a (not conditioned) Markov chain with a different set of transition probabilities. We consider a few classical examples and discuss some properties of this construction. Then, we work with Doob's transform of simple random walk in two dimensions with respect to its potential kernel. It turns out that this conditioned simple random walk is a fascinating object in its own right: just to cite one of its properties, the probability that a site y is ever visited by a walk started somewhere close to the origin converges to $1/2$ as $y \to \infty$. Perhaps even more surprisingly, the proportion of visited sites of "typical" large sets approaches in distribution a Uniform[0, 1] random variable.
5. *Intermezzo: soft local times and Poisson processes of objects.* This chapter is about two subjects, apparently unrelated to simple random walk.

One is called soft local times; generally speaking, the method of soft local times is a way to construct an adapted stochastic process on a general space Σ using an auxiliary Poisson point process on $\Sigma \times \mathbb{R}_+$. In Chapter 6, this method will be an important tool for dealing with *excursion processes*. Another topic we discuss is "Poisson processes of infinite objects", using as an introductory example the Poisson line process. While this example per se is not formally necessary for the book, it helps us to build some intuition about what will happen in the next chapter.

6. *Random interlacements.* In this chapter, we discuss random interlacements, which are Poisson processes of simple random walk trajectories. First, we review Sznitman's random interlacements model [93] in dimension $d \geq 3$. Then we discuss the two-dimensional case recently introduced in [26]; it is here that various plot lines of this book finally meet. This model will be built of the trajectories of simple random walk conditioned on not hitting the origin, studied in Chapter 4. Using the estimates of two-dimensional capacities and hitting probabilities obtained with the technique of Chapters 3 and 4, we then prove several properties of the model, and the soft local times of Chapter 5 will enter as an important tool in some of these proofs. As stated by Sznitman in [97], "One has good decoupling properties of the excursions ... when the boxes are sufficiently far apart. The soft local time technique ... offers a very convenient tool to express these properties".

Acknowledgements

Special thanks are due to those whose collaborations directly relate to material presented in one or more of the chapters of this book: Francis Comets, Nina Gantert, Mikhail Menshikov, Augusto Teixeira, Marina Vachkovskaia, and Andrew Wade. I also thank Daniel Ungaretti Borges, Darcy Camargo, Caio Alves, Gustavo Henrique Tasca, Victor Freguglia Souza, Matheus Gorito de Paula, and Thainá Soares Silva, who read the manuscript at different stages and made many useful comments and suggestions.

Most of this work was done when the author held a professor position at the University of Campinas (UNICAMP), Brazil. The author takes this occasion to express his profound gratitude to all his Brazilian friends and colleagues.

Notation

Here we list the notation recurrently used in this book.

- We write $X := \ldots$ to indicate the definition of X, and will also occasionally use $\ldots =: X$.
- $a \wedge b := \min\{a, b\}$, $a \vee b := \max\{a, b\}$.
- For a real number x, $\lfloor x \rfloor$ is the largest integer not exceeding x, and $\lceil x \rceil$ is the smallest integer no less than x.

Sets

- $|A|$ is the cardinality of a finite set A.
- $\mathbb{R}$ is the set of real numbers, and $\mathbb{R}_+ = [0, +\infty)$ is the set of real nonnegative numbers.
- $\mathbb{Z}$ is the set of integer numbers, $\mathbb{N} = \{1, 2, 3, \ldots\}$ is the set of natural numbers, $\mathbb{Z}_+ = \{0, 1, 2, \ldots\}$ is the set of integer nonnegative numbers, $\overline{\mathbb{Z}}_+ = \mathbb{Z}_+ \cup \{+\infty\}$.
- $\mathbb{R}^d$ is the d-dimensional Euclidean space and $\mathbb{Z}^d$ is the d-dimensional integer lattice (with the usual graph structure).
- $\mathbb{Z}_n^d = \mathbb{Z}^d / n\mathbb{Z}^d$ is the d-dimensional torus of (linear) size n (with the graph structure inherited from $\mathbb{Z}^d$).
- For $A \subset \mathbb{Z}^d$, $A^{\complement} = \mathbb{Z}^d \setminus A$ is the complement of A, $\partial A = \{x \in A : \text{there exists } y \in A^{\complement} \text{ such that } x \sim y\}$ is the boundary of A, and $\partial_e A = \partial(A^{\complement})$ is the external boundary of A.
- $\mathcal{N} = \partial_e\{0\} = \{\pm e_{1,2}\} \subset \mathbb{Z}^2$ is the set of the four neighbours of the origin (in two dimensions).
- $\mathrm{B}(x, r) = \{y : \|y - x\| \leq r\}$ is the ball (disk) in $\mathbb{R}^d$ or $\mathbb{Z}^d$; $\mathrm{B}(r)$ stands for $\mathrm{B}(0, r)$.

11

Asymptotics of functions

- $f(x) \asymp g(x)$ means that there exist $0 < C_1 < C_2 < \infty$ such that $C_1 g(x) \leq f(x) \leq C_2 g(x)$ for all x; $f(x) \lesssim g(x)$ means that there is $C_3 > 0$ such that $f(x) \leq C_3 g(x)$ for all x.
- $f(x) = O(g(x))$ as $x \to a$ means that $\limsup_{x \to a} \left| \frac{f(x)}{g(x)} \right| < \infty$, where $a \in \mathbb{R} \cup \{\infty\}$; $f(x) = o(g(x))$ as $x \to a$ means that $\lim_{x \to a} \frac{f(x)}{g(x)} = 0$.

Euclidean spaces and vectors

- $\|x\|$ is the Euclidean norm of $x \in \mathbb{R}^d$ or $x \in \mathbb{Z}^d$.
- $x \cdot y$ is the scalar product of $x, y \in \mathbb{R}^d$.
- We write $x \sim y$ if x and y are neighbours in $\mathbb{Z}^d$ (i.e., $x, y \in \mathbb{Z}^d$ and $\|x - y\| = 1$).
- $(e_k, k = 1, \ldots, d)$ are the canonical coordinate vectors in $\mathbb{R}^d$ or $\mathbb{Z}^d$.
- For $A, B \subset \mathbb{R}^d$ or $\mathbb{Z}^d$, $\text{dist}(A, B) = \inf_{x \in A, y \in B} \|x - y\|$, and $\text{dist}(x, A) := \text{dist}(\{x\}, A)$; also, $\text{diam}(A) = \sup_{x, y \in A} \|x - y\|$.

General probability and stochastic processes

- $(\mathcal{F}_n, n \geq 0)$ is a filtration (a nondecreasing sequence of sigma-algebras).
- a.s. stands for "almost surely" (with probability 1).
- $\mathbf{1}\{event\}$ is the indicator function of event $\{event\}$.
- $(p(x, y), x, y \in \Sigma)$ are transition probabilities of a Markov chain on a state space Σ, and $(p_n(x, y), x, y \in \Sigma)$ are the n-step transition probabilities.
- $\mathbb{P}_x$ and $\mathbb{E}_x$ are probability and expectation for a process (normally, a random walk – the one that we are considering at the moment) starting from x.
- SRW is an abbreviation for "simple random walk".
- $L_n(z) = \sum_{k=1}^{n} \mathbf{1}\{X_k = z\}$ is the local time at z of the process X at time n; we write $L_n^X(z)$ in case when there might be an ambiguity about which process we are considering.[3]
- $\mathcal{G}_n(z)$ is the soft local time of the process at time n at site z.

Simple random walk

- $(S_n, n \geq 0)$ is the simple random walk in $\mathbb{Z}^d$.
- $\tau_A \geq 0$ and $\tau_A^+ \geq 1$ are entrance and hitting times of A by the SRW.
- $\text{Es}_A(x) = \mathbb{P}_x[\tau_A^+ = \infty]\mathbf{1}\{x \in A\}$ is the escape probability from $x \in A$ for SRW in dimensions $d \geq 3$.

---

[3] We use different notation for local times of SRW and conditioned SRW; see the following notations for (conditioned) simple random walk.

- $G(\cdot, \cdot)$ is Green's function for the SRW in three or more dimensions, $G_\Lambda(\cdot, \cdot)$ is Green's function restricted on Λ.
- $a(\cdot)$ is the potential kernel for the two-dimensional SRW.
- $hm_A(x)$ is the value of the harmonic measure in $x \in A$.
- $cap(A)$ is the capacity of A (in two or more dimensions).
- $N_x = \sum_{j=0}^\infty \mathbf{1}\{S_j = x\}$ is the total number of visits to x, and $N_x^{(k)} = \sum_{j=0}^k \mathbf{1}\{S_k = x\}$ is the total number of visits to x up to time k (i.e., the local time at x at time k).
- Let A be a fixed subset of $\mathbb{Z}^2$; then $N_x^\flat = \sum_{j=0}^{\tau_A^+ - 1} \mathbf{1}\{S_j = x\}$ is the number of visits to x *before* the first return to A, and by $N_x^\sharp = \sum_{j=\tau_A^+}^\infty \mathbf{1}\{S_j = x\}$ the number of visits to x *after* the first return to A (with $N_x^\sharp = 0$ on $\{\tau_A^+ = \infty\}$).

Conditioned simple random walk

- $(\widehat{S}_n, n \geq 0)$ is the simple random walk in $\mathbb{Z}^2$ conditioned on never hitting the origin.
- $\hat{\tau}_A \geq 0$ and $\hat{\tau}_A^+ \geq 1$ are entrance and hitting times of A by the conditioned SRW.
- $\widehat{Es}_A(x) = \mathbb{P}_x[\hat{\tau}_A^+ = \infty]\mathbf{1}\{x \in A\}$ is the escape probability from $x \in A$ for the conditioned SRW in two dimensions.
- $\widehat{hm}_A(x)$ is the harmonic measure for the conditioned walk.
- $\widehat{cap}(A) = cap(A \cup \{0\})$ is the capacity of set A with respect to the conditioned SRW.
- $\widehat{G}(x, y) = a(y)(a(x) + a(y) - a(x - y))/a(x)$ is Green's function of the conditioned walk.
- $\hat{\ell}(x, y) = 1 + \frac{a(y) - a(x-y)}{a(x)} = \frac{\widehat{G}(x,y)}{a(y)}$, and $\hat{g}(x, y) = \frac{\widehat{G}(x,y)}{a^2(y)} = \hat{g}(y, x)$ is the "symmetrized" $\widehat{G}$.
- $\widehat{N}_x, \widehat{N}_x^{(k)}, \widehat{N}_x^\flat, \widehat{N}_x^\sharp$ are defined just as $N_x, N_x^{(k)}, N_x^\flat, N_x^\sharp$, but with the conditioned walk $\widehat{S}$ instead of SRW S.

Random interlacements

- $RI(\alpha)$ is the random interlacement process on level $\alpha > 0$.
- $\mathbb{E}^u$ is the expectation for random interlacements on level $u > 0$.
- I^u and V^u are the interlacement and vacant sets for the random interlacement model on level $u > 0$ in dimensions $d \geq 3$; in two dimensions, we usually denote the level by α, so these become I^α and V^α.

1

Introduction

The main subject of this book is simple random walk (also abbreviated as SRW) on the integer lattice $\mathbb{Z}^d$ and we will pay special attention to the case $d = 2$. SRW is a discrete-time stochastic process which is defined in the following way: if at a given time the walker is at $x \in \mathbb{Z}^d$, then at the next time moment it will be at one of x's $2d$ neighbours chosen uniformly at random.[1] In other words, the probability that the walk follows a fixed length-n path of nearest-neighbour sites equals $(2d)^{-n}$. As a general fact, a random walk may be recurrent (i.e., almost surely it returns infinitely many times to its starting location) or transient (i.e., with positive probability it never returns to its starting location). A fundamental result about SRWs on integer lattices is Pólya's classical theorem [76]:

Theorem 1.1. *Simple random walk in dimension d is recurrent for $d = 1, 2$ and transient for $d \geq 3$.*

A well-known interpretation of this fact, attributed to Shizuo Kakutani, is: "a drunken man always returns home, but a drunken bird will eventually be lost". This observation may explain why birds do not drink vodka. Still, despite recurrence, the drunken man's life is not so easy either: as we will see, it may take him *quite* some time to return home.

Indeed, as we will see in (3.42), the probability that two-dimensional SRW gets more than distance n away from its starting position without revisiting it is approximately $(1.0293737 + \frac{2}{\pi} \ln n)^{-1}$ (and this formula becomes *very* precise as n grows). While this probability indeed converges to zero as $n \to \infty$, it is important to notice how slow this convergence is. To present a couple of concrete examples, assume that the size of the walker's step is equal to 1 metre. First of all, let us go to one of the most beautiful cities in the world, Paris, and start walking from its centre. The

---

[1] Here, the author had to resist the temptation of putting a picture of an SRW's trajectory in view of the huge number of animated versions easily available in the Internet, e.g., at https://en.wikipedia.org/wiki/Random_walk.

radius of Paris is around 5000m, and $(1.0293737 + \frac{2}{\pi} \ln 5000)^{-1}$ is approximately 0.155; that is, in roughly one occasion out of seven you would come to the Boulevard Périphérique before returning to your starting location. The next example is a bit more extreme: let us do the same walk on the *galactic plane* of our galaxy. (Yes, when one starts in the centre of our galaxy, there is a risk that the starting location could happen to be too close to a massive black hole;[2] we restrict ourselves to purely mathematical aspects of the preceding question, though.) The radius of the Milky Way galaxy is around 10^{21}m, and $(1.0293737 + \frac{2}{\pi} \ln 10^{21})^{-1} \approx 0.031$, which is *surprisingly* large. Indeed, this means that the walker[3] would revisit the origin only around 30 times on average, before leaving the galaxy; this is not something one would normally expect from a *recurrent* process.

Incidentally, these sorts of facts explain why it is difficult to verify conjectures about two-dimensional SRW using computer simulations. (For example, imagine that one needs to estimate how long we will wait until the walk returns to the origin, say, a hundred times.)

As we will see in Section 2.1, the recurrence of d-dimensional SRW is related to the divergence of the series $\sum_{n=1}^{\infty} n^{-d/2}$. Notice that this series diverges if and only if $d \leq 2$, and for $d = 2$ it is the harmonic series that diverges quite slowly. This might explain why the two-dimensional case is, in some sense, *really* critical (and therefore gives rise to the previous "strange" examples). It is always interesting to study critical cases – they frequently exhibit behaviours not observable away from criticality. For this reason, in this book we dedicate more attention to dimension two than to other dimensions: two-dimensional SRW is a fascinating mathematical object indeed and this already justifies one's interest in exploring its properties (and also permits the author to keep this introduction short).

The next section is intentionally kept concise, since it is not really intended for *reading* but rather for occasional use as a reference.

1.1 Markov chains and martingales: basic definitions and facts

First, let us recall some basic definitions related to real-valued stochastic processes in discrete time. In the following, all random variables are defined on a common probability space $(\Omega, \mathcal{F}, \mathbb{P})$. We write $\mathbb{E}$ for expectation corresponding to $\mathbb{P}$, which will be applied to real-valued random variables. Set $\mathbb{N} = \{1, 2, 3, \ldots\}, \mathbb{Z}_+ = \{0, 1, 2, \ldots\}, \overline{\mathbb{Z}}_+ = \mathbb{Z}_+ \cup \{+\infty\}$.

---

[2] https://en.wikipedia.org/wiki/Sagittarius_A*.
[3] Given the circumstances, let me not say "you" here.

Definition 1.2 (Basic concepts for discrete-time stochastic processes).

- A discrete-time real-valued *stochastic process* is a sequence of random variables $X_n : (\Omega, \mathcal{F}) \to (\mathbb{R}, \mathcal{B})$ indexed by $n \in \mathbb{Z}_+$, where $\mathcal{B}$ is the Borel σ-field. We write such sequences as $(X_n, n \geq 0)$, with the understanding that the time index n is always an integer.
- A *filtration* is a sequence of σ-fields $(\mathcal{F}_n, n \geq 0)$ such that $\mathcal{F}_n \subset \mathcal{F}_{n+1} \subset \mathcal{F}$ for all $n \geq 0$. Let us also define $\mathcal{F}_\infty := \sigma(\bigcup_{n \geq 0} \mathcal{F}_n) \subset \mathcal{F}$.
- A stochastic process $(X_n, n \geq 0)$ is *adapted* to a filtration $(\mathcal{F}_n, n \geq 0)$ if X_n is $\mathcal{F}_n$-measurable for all $n \in \mathbb{Z}_+$.
- For a (possibly infinite) random variable $\tau \in \overline{\mathbb{Z}}_+$, the random variable X_τ is (as the notation suggests) equal to X_n on $\{\tau = n\}$ for finite $n \in \mathbb{Z}_+$ and equal to $X_\infty := \limsup_{n \to \infty} X_n$ on $\{\tau = \infty\}$.
- A (possibly infinite) random variable $\tau \in \overline{\mathbb{Z}}_+$ is a *stopping time* with respect to a filtration $(\mathcal{F}_n, n \geq 0)$ if $\{\tau = n\} \in \mathcal{F}_n$ for all $n \geq 0$.
- If τ is a stopping time, the corresponding σ-field $\mathcal{F}_\tau$ consists of all events $A \in \mathcal{F}_\infty$ such that $A \cap \{\tau \leq n\} \in \mathcal{F}_n$ for all $n \in \overline{\mathbb{Z}}_+$. Note that $\mathcal{F}_\tau \subset \mathcal{F}_\infty$; events in $\mathcal{F}_\tau$ include $\{\tau = \infty\}$, as well as $\{X_\tau \in B\}$ for all $B \in \mathcal{B}$.
- For $A \in \mathcal{B}$, let us define

$$\tau_A = \min\{n \geq 0 : X_n \in A\}, \tag{1.1}$$

and

$$\tau_A^+ = \min\{n \geq 1 : X_n \in A\}; \tag{1.2}$$

we may refer to either τ_A or τ_A^+ as the *hitting time* of A (also called the *passage time* into A). It is straightforward to check that both τ_A and τ_A^+ are stopping times.

Observe that, for any stochastic process $(X_n, n \geq 0)$, it is possible to define the minimal filtration to which this process is adapted via $\mathcal{F}_n = \sigma(X_0, X_1, \ldots, X_n)$. This is the so-called *natural filtration*.

To keep the notation concise, we will frequently write X_n and $\mathcal{F}_n$ instead of $(X_n, n \geq 0)$ and $(\mathcal{F}_n, n \geq 0)$ and so on, when no confusion will arise.

Next, we need to recall some martingale-related definitions and facts.

Definition 1.3 (Martingales, submartingales, supermartingales). A real-valued stochastic process X_n adapted to a filtration $\mathcal{F}_n$ is a *martingale* (with respect to $\mathcal{F}_n$) if, for all $n \geq 0$,

(i) $\mathbb{E}|X_n| < \infty$, and
(ii) $\mathbb{E}[X_{n+1} - X_n \mid \mathcal{F}_n] = 0$.

If in (ii) "=" is replaced by "≥" (respectively, "≤"), then X_n is called a *submartingale* (respectively, *supermartingale*). If the filtration is not specified, that means that the natural filtration is used.

Clearly, if X_n is a submartingale, then $(-X_n)$ is a supermartingale, and vice versa; a martingale is both a submartingale and a supermartingale. Also, it is elementary to observe that if X_n is a (sub-, super-)martingale, then so is $X_{n \wedge \tau}$ for any stopping time τ.

Martingales have a number of remarkable properties, which we will not even try to elaborate on here. Let us only cite the paper [75], whose title speaks for itself. In the following, we mention only the results needed in this book.

We start with

Theorem 1.4 (Martingale convergence theorem). *Assume that X_n is a submartingale such that $\sup_n \mathbb{E} X_n^+ < \infty$. Then there is an integrable random variable X such that $X_n \to X$ a.s. as $n \to \infty$.*

Observe that, under the hypotheses of Theorem 1.4, the sequence $\mathbb{E} X_n$ is non-decreasing (by the submartingale property) and bounded above by $\sup_n \mathbb{E}[X_n^+]$; so $\lim_{n \to \infty} \mathbb{E} X_n$ exists and is finite. However, it is not necessarily equal to $\mathbb{E} X$.

Using Theorem 1.4 and Fatou's lemma, it is straightforward to obtain the following result.

Theorem 1.5 (Convergence of non-negative supermartingales). *Assume that $X_n \geq 0$ is a supermartingale. Then there is an integrable random variable X such that $X_n \to X$ a.s. as $n \to \infty$, and $\mathbb{E} X \leq \mathbb{E} X_0$.*

Another fundamental result that we will use frequently is the following:

Theorem 1.6 (Optional stopping theorem). *Suppose that $\sigma \leq \tau$ are stopping times, and $X_{\tau \wedge n}$ is a uniformly integrable submartingale. Then $\mathbb{E} X_\sigma \leq \mathbb{E} X_\tau < \infty$ and $X_\sigma \leq \mathbb{E}[X_\tau \mid \mathcal{F}_\sigma]$ a.s.*

Note that, if X_n is a uniformly integrable submartingale and τ is any stopping time, then it can be shown that $X_{\tau \wedge n}$ is also uniformly integrable: see, e.g., section 5.7 of [44]. Also, observe that two applications of Theorem 1.6, one with $\sigma = 0$ and one with $\tau = \infty$, show that for any uniformly integrable submartingale X_n and any stopping time τ, it holds that $\mathbb{E} X_0 \leq \mathbb{E} X_\tau \leq \mathbb{E} X_\infty < \infty$, where $X_\infty := \limsup_{n \to \infty} X_n = \lim_{n \to \infty} X_n$ exists and is integrable, by Theorem 1.4.

Theorem 1.6 has the following corollary, obtained by considering $\sigma = 0$

and using well-known sufficient conditions for uniform integrability (e.g., sections 4.5 and 4.7 of [44]).

Corollary 1.7. *Let X_n be a submartingale and τ a finite stopping time. For a constant $c > 0$, suppose that at least one of the following conditions holds:*

(i) $\tau \leq c$ a.s.;
(ii) $|X_{n \wedge \tau}| \leq c$ a.s. for all $n \geq 0$;
(iii) $\mathbb{E}\tau < \infty$ and $\mathbb{E}[|X_{n+1} - X_n| \mid \mathcal{F}_n] \leq c$ a.s. for all $n \geq 0$.

Then $\mathbb{E}X_\tau \geq \mathbb{E}X_0$. If X_n is a martingale and at least one of the conditions (i) through (iii) holds, then $\mathbb{E}X_\tau = \mathbb{E}X_0$.

Next, we recall some fundamental definitions and facts for Markov processes in discrete time and with countable state space, also known as *countable Markov chains*. In the following, $(X_n, n \geq 0)$ is a sequence of random variables taking values on a countable set Σ.

Definition 1.8 (Markov chains).

- A process X_n is a *Markov chain* if, for any $y \in \Sigma$, any $n \geq 0$, and any $m \geq 1$,

$$\mathbb{P}[X_{n+m} = y \mid X_0, \ldots, X_n] = \mathbb{P}[X_{n+m} = y \mid X_n], \quad \text{a.s..} \qquad (1.3)$$

  This is the *Markov property*.
- If there is no dependence on n in (1.3), the Markov chain is *homogeneous in time* (or *time homogeneous*). Unless explicitly stated otherwise, all Markov chains considered in this book are assumed to be time homogeneous. In this case, the Markov property (1.3) becomes

$$\mathbb{P}[X_{n+m} = y \mid \mathcal{F}_n] = p_m(X_n, y), \quad \text{a.s.,} \qquad (1.4)$$

  where $p_m : \Sigma \times \Sigma \to [0, 1]$ are the m-step Markov *transition probabilities*, for which the Chapman–Kolmogorov equation holds: $p_{n+m}(x, y) = \sum_{z \in \Sigma} p_n(x, z) p_m(z, y)$. Also, we write $p(x, y) := \mathbb{P}[X_1 = y \mid X_0 = x] = p_1(x, y)$ for the one-step transition probabilities of the Markov chain.
- We use the shorthand notation $\mathbb{P}_x[\cdot] = \mathbb{P}[\cdot \mid X_0 = x]$ and $\mathbb{E}_x[\cdot] = \mathbb{E}[\cdot \mid X_0 = x]$ for probability and expectation for the time homogeneous Markov chain starting from initial state $x \in \Sigma$.
- A time homogeneous, countable Markov chain is *irreducible* if for all $x, y \in \Sigma$ there exists $n_0 = n_0(x, y) \geq 1$ such that $p_{n_0}(x, y) > 0$.

- For an irreducible Markov chain, we define its *period* as the greatest common divisor of $\{n \geq 1 : p_n(x,x) > 0\}$ (it is not difficult to show that it does not depend on the choice of $x \in \Sigma$). An irreducible Markov chain with period 1 is called *aperiodic*.
- Let X_n be a Markov chain and τ be a stopping time with respect to the natural filtration of X_n. Then for all $x, y_1, \ldots, y_k \in \Sigma$, $n_1, \ldots, n_k \geq 1$, it holds that

$$\mathbb{P}[X_{\tau+n_j} = y_j, j = 1, \ldots, k \mid \mathcal{F}_\tau, X_\tau = x] = \mathbb{P}_x[X_{\tau+n_j} = y_j, j = 1, \ldots, k]$$

(this is the *strong Markov property*).
- For a Markov chain, a probability measure $(\pi(x), x \in \Sigma)$ is called an *invariant measure* if $\sum_{x \in \Sigma} \pi(x) p(x,y) = \pi(y)$ for all $y \in \Sigma$. It then holds that $\mathbb{P}_\pi[X_n = y] = \pi(y)$ for all n and y (where $\mathbb{P}_\pi$ means that the initial state of the process is chosen according to π).

Suppose now that X_n is a countable Markov chain. Recall the definitions of hitting times τ_A and τ_A^+ from (1.1)–(1.2). For $x \in \Sigma$, we use the notation $\tau_x^+ := \tau_{\{x\}}^+$ and $\tau_x := \tau_{\{x\}}$ for hitting times of one-point sets. Note that for any $x \in A$ it holds that $\mathbb{P}_x[\tau_A = 0] = 1$, while $\tau_A^+ \geq 1$ is then the *return time* to A. Also note that $\mathbb{P}_x[\tau_A = \tau_A^+] = 1$ for all $x \in \Sigma \setminus A$.

Definition 1.9. For a countable Markov chain X_n, a state $x \in \Sigma$ is called

- *recurrent* if $\mathbb{P}_x[\tau_x^+ < \infty] = 1$;
- *transient* if $\mathbb{P}_x[\tau_x^+ < \infty] < 1$.

A recurrent state x is classified further as

- *positive recurrent* if $\mathbb{E}_x \tau_x^+ < \infty$;
- *null recurrent* if $\mathbb{E}_x \tau_x^+ = \infty$.

It is straightforward to see that the four properties in Definition 1.9 are *class properties*, which entails the following statement.

Proposition 1.10. *For an irreducible Markov chain, if a state $x \in \Sigma$ is recurrent (respectively, positive recurrent, null recurrent, transient), then all states in Σ are recurrent (respectively, positive recurrent, null recurrent, transient).*

By the preceding fact, it is legitimate to call an irreducible Markov chain itself recurrent (positive recurrent, null recurrent, transient).

Next, the following proposition is an easy consequence of the strong Markov property.

Proposition 1.11. *For an irreducible Markov chain, if a state $x \in \Sigma$ is recurrent (respectively, transient), then, regardless of the initial position of the process, it will be visited infinitely (respectively, finitely) many times almost surely.*

Finally, let us state the following simple result which sometimes helps in proving recurrence or transience of Markov chains.

Lemma 1.12. *Let X_n be an irreducible Markov chain on a countable state space Σ.*

(i) *If for some $x \in \Sigma$ and some nonempty $A \subset \Sigma$ it holds that $\mathbb{P}_x[\tau_A < \infty] < 1$, then X_n is transient.*

(ii) *If for some finite nonempty $A \subset \Sigma$ and all $x \in \Sigma \setminus A$ it holds that $\mathbb{P}_x[\tau_A < \infty] = 1$, then X_n is recurrent.*

(For the proof, cf. e.g. lemma 2.5.1 of [71].)

2

Recurrence of two-dimensional simple random walk

This chapter is mainly devoted to the proof of the recurrence part of Theorem 1.1 (although we still discuss the transience in higher dimensions later in the exercises). We first present a direct "path-counting" proof, and then discuss the well-known correspondence between reversible Markov chains and electrical networks, which also yields a beautiful proof of recurrence of SRW in dimensions one and two. Then, we go for a side-quest: we do a basic exploration of the Lyapunov function method, a powerful tool for proving recurrence or transience of general Markov chains. With this method, we add yet another proof of recurrence of two-dimensional SRW to our collection.

2.1 Classical proof

In this section, we present the classical combinatorial proof of recurrence of two-dimensional simple random walk.

Let us start with some general observations on recurrence and transience of random walks, which, in fact, are valid in a much broader setting. Namely, we will prove that the number of visits to the origin is a.s. finite if and only if the *expected* number of visits to the origin is finite (note that this is something which is *not* true for general random variables). This is a useful fact, because, as it frequently happens, it is easier to control the expectation than the random variable itself.

Let $p_m(x, y) = \mathbb{P}_x[S_m = y]$ be the transition probability from x to y in m steps for the simple random walk in d dimensions. Let $q_d = \mathbb{P}_0[\tau_0^+ < \infty]$ be the probability that, starting at the origin, the walk eventually returns to the origin. If $q_d < 1$, then the total number of visits (counting the initial instance $S_0 = 0$ as a visit) is a geometric random variable with success probability $1 - q_d$, which has expectation $(1 - q_d)^{-1} < \infty$. If $q_d = 1$, then, clearly, the walk visits the origin infinitely many times a.s.. So, random walk is transient (i.e., $q_d < 1$) if and only if the *expected* number of visits to the origin is finite. This expected number equals[1]

$$\mathbb{E}_0 \sum_{k=0}^{\infty} \mathbf{1}\{S_k = 0\} = \sum_{k=0}^{\infty} \mathbb{E}_0 \mathbf{1}\{S_k = 0\} = \sum_{n=0}^{\infty} \mathbb{P}_0[S_{2n} = 0]$$

(observe that the walk can be at the starting point only after an even number of steps). We thus obtain that the recurrence of the walk is *equivalent* to

$$\sum_{n=0}^{\infty} p_{2n}(0,0) = \infty. \tag{2.1}$$

Before actually proving anything, let us try to *understand* why Theorem 1.1 should hold. One can represent the d-dimensional simple random walk S as

$$S_n = X_1 + \cdots + X_n,$$

where $(X_k, k \geq 1)$ are independent and identically distributed (i.i.d.) random vectors, uniformly distributed on the set $\{\pm e_j, j = 1, \ldots, d\}$, where $e_1, \ldots, e_d$ is the canonical basis of $\mathbb{R}^d$. Since these random vectors are centred (expectation is equal to 0, component-wise), one can apply the (multivariate) Central Limit Theorem (CLT) to obtain that $S_n / \sqrt{n}$ converges in distribution to a (multivariate) centred Normal random vector with a diagonal covariance matrix. That is, it is reasonable to expect that S_n should be at distance of order $\sqrt{n}$ from the origin.

So, what about $p_{2n}(0,0)$? Well, if $x, y \in \mathbb{Z}^d$ are two *even* sites[2] at distance of order at most $\sqrt{n}$ from the origin, then our CLT intuition tell us that $p_{2n}(0, x)$ and $p_{2n}(0, y)$ should be *comparable*, i.e., their ratio should be bounded away from 0 and ∞. In fact, this statement can be made rigorous by using the *local* Central Limit Theorem (e.g., theorem 2.1.1 from [63]). Now, if there are $O(n^{d/2})$ sites where $p_{2n}(0, \cdot)$ are comparable, then the value of these probabilities (including $p_{2n}(0,0)$) should be of order $n^{-d/2}$. It remains only to observe that $\sum_{n=1}^{\infty} n^{-d/2}$ diverges only for $d = 1$ and 2 to convince oneself that Pólya's theorem indeed holds. Notice, by the way, that for $d = 2$ we have the harmonic series which diverges just barely; its partial sums have only logarithmic growth.[3]

Now, let us *prove* that (2.1) holds for one- and two-dimensional simple random walks. In the one-dimensional case, it is quite simple to calculate $p_{2n}(0,0)$: it is the probability that a Binomial$(2n, \frac{1}{2})$-random variable

---

[1] Note that we can put the expectation inside the sum because of the Monotone Convergence Theorem.

[2] A site is called even if the sum of its coordinates is even; observe that the origin is even.

[3] As some physicists say, "in practice, logarithm is a constant!"

equals 0, so it is $2^{-2n}\binom{2n}{n}$. Certainly, this expression is concise and beautiful; it is, however, not a priori clear which asymptotic behaviour it has (as it frequently happens with concise and beautiful formulas). To clarify this, we use Stirling's approximation[4], $n! = \sqrt{2\pi n}(n/e)^n(1+o(1))$, to obtain that

$$2^{-2n}\binom{2n}{n} = 2^{-2n}\frac{(2n)!}{(n!)^2}$$

$$= 2^{-2n}\frac{\sqrt{4\pi n}(2n/e)^{2n}}{2\pi n(n/e)^{2n}}(1+o(1))$$

(fortunately, almost everything cancels)

$$= \frac{1}{\sqrt{\pi n}}(1+o(1)). \tag{2.2}$$

The series $\sum_{k=1}^{\infty} k^{-1/2}$ diverges, so (2.1) holds, and this implies recurrence in dimension 1.

Let us now deal with the two-dimensional case. For this, we first count the number of paths N_{2n} of length $2n$ that start and end at the origin. For such a path, the number of steps up must be equal to the number of steps down, and the number of steps to the right must be equal to the number of steps to the left. The total number of steps up (and, also, down) can be any integer k between 0 and n; in this case, the trajectory must have $n-k$ steps to the left and $n-k$ steps to the right. So, if the number of steps up is k, the total number of trajectories starting and ending at the origin is the polynomial coefficient $\binom{2n}{k,k,n-k,n-k}$. This means that

$$N_{2n} = \sum_{k=0}^{n}\binom{2n}{k,k,n-k,n-k} = \sum_{k=0}^{n}\frac{(2n)!}{(k!)^2((n-k)!)^2}.$$

Note that

$$\frac{(2n)!}{(k!)^2((n-k)!)^2} = \binom{2n}{n}\binom{n}{k}\binom{n}{n-k};$$

the last two factors are clearly equal, but in a few lines it will become clear why we have chosen to write it this way. Since the probability of any particular trajectory of length m is 4^{-m}, we have

$$p_{2n}(0,0) = 4^{-2n}N_{2n}$$

$$= 4^{-2n}\binom{2n}{n}\sum_{k=0}^{n}\binom{n}{k}\binom{n}{n-k}. \tag{2.3}$$

---

[4] See, e.g., http://mathworld.wolfram.com/StirlingsApproximation.html.

There is a nice combinatorial argument that allows one to deal with the sum in the right-hand side of (2.3). Consider a group of $2n$ children of which n are boys and n are girls. What is the number of ways to choose a subgroup of n children from that group? On one hand, since there are no restrictions on the gender composition of the subgroup, the answer is simply $\binom{2n}{n}$. On the other hand, the number of boys in the subgroup can vary from 0 to n, and, given that there are k boys (and, therefore, $n-k$ girls), there are $\binom{n}{k}\binom{n}{n-k}$ ways to choose the subgroup; so, the answer is precisely the preceding sum. This means that this sum just equals $\binom{2n}{n}$, and we thus obtain that, in two dimensions,

$$p_{2n}(0,0) = \left(2^{-2n}\binom{2n}{n}\right)^2. \tag{2.4}$$

The calculation (2.2) then implies that

$$p_{2n}(0,0) = \frac{1}{\pi n}(1 + o(1)) \tag{2.5}$$

for two-dimensional SRW, and, using the fact that the harmonic series diverges, we obtain (2.1) and therefore recurrence. □

It is curious to notice that (2.4) means that the probability of being at the origin at time $2n$ for two-dimensional SRW is *exactly* the square of corresponding probability in one dimension. Such coincidences usually happen for a reason, and this case is no exception: in fact, it is possible to *decouple* the "one-dimensional components" of the two-dimensional SRW by considering its projections on the axes rotated $\pi/4$ anticlockwise; these projections are *independent*.[5] Indeed, it is straightforward to verify[6] that $S_n \cdot (e_1 + e_2)$ and $S_n \cdot (e_1 - e_2)$ are independent one-dimensional SRWs.

2.2 Electrical networks

First, we recall the following.

Definition 2.1. A Markov chain with transition probabilities $(p(x, y), x, y \in \Sigma)$ is called *reversible* with the reversible measure $\pi = (\pi(x), x \in \Sigma)$ when $\pi(x)p(x, y) = \pi(y)p(y, x)$ for all $x, y \in \Sigma$.

It is important to note that, in Definition 2.1, we do not assume that π is a *probability* measure; its total mass can be any positive number or even

---

[5] This fact is folklore, but the author thanks Alejandro Ramirez for making him aware of it.

[6] This is left as an exercise.

infinity (in the case when the state space Σ is infinite). For example, note that $\pi \equiv 1$ is reversible for the SRW (in any dimension). However, when the total mass of the reversible measure is 1, it is also invariant for the Markov chain, as shown by the following simple calculation:

$$\sum_{x \in \Sigma} \pi(x)p(x, y) = \sum_{x \in \Sigma} \pi(y)p(y, x) = \pi(y) \sum_{x \in \Sigma} p(y, x) = \pi(y).$$

Also, a reversible measure cannot be unique: if π is a reversible measure, then so is $c\pi$ for any $c > 0$.

In this book we will not take a deep dive into the theory of reversible Markov chains; let us only note that the latter is rich and useful. It is good to know the following *criterion* of reversibility: one can check if the Markov chain is reversible without actually calculating the reversible measure.

Theorem 2.2. *A Markov chain is reversible if and only if for any cycle* $x_0, x_1, \ldots, x_{n-1}, x_n = x_0$ *of states it holds that:*

$$\prod_{k=0}^{n-1} p(x_k, x_{k+1}) = \prod_{k=1}^{n} p(x_k, x_{k-1}); \tag{2.6}$$

that is, the product of the transition probabilities along the cycle does not depend on the direction.

Proof First, assume that the Markov chain is reversible, and let us do the proof for $n = 2$ (i.e., cycle of size 3); the reader will easily see that the same argument works for all n. The idea is to multiply (2.6) by $\pi(x_0)$ and then "let it go through the cycle":

$$\pi(x_0)p(x_0, x_1)p(x_1, x_2)p(x_2, x_0) = p(x_1, x_0)\pi(x_1)p(x_1, x_2)p(x_2, x_0)$$
$$= p(x_1, x_0)p(x_2, x_1)\pi(x_2)p(x_2, x_0)$$
$$= p(x_1, x_0)p(x_2, x_1)p(x_0, x_2)\pi(x_0),$$

then we cancel $\pi(x_0)$ and obtain the claim.

Now, let us assume that (2.6) holds, and prove that the Markov chain is reversible. The main difficulty is to find a good candidate for the reversible measure – in principle, a priori there is none. But if one recalls what the reversible measure for one-dimensional nearest-neighbour random walk looks like (see Exercise 2.7), one may come out with following procedure:

- fix some site x_0 and put $\pi(x_0) = 1$;

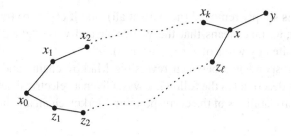

Figure 2.1 On the proof of the reversibility criterion.

- for $x \neq x_0$, fix a path $(x_0, x_1, \ldots, x_k, x)$ which goes from x_0 to x; then, set

$$\pi(x) = \frac{p(x_0, x_1)p(x_1, x_2) \cdots p(x_k, x)}{p(x_1, x_0)p(x_2, x_1) \cdots p(x, x_k)}. \qquad (2.7)$$

We then need to check that this definition is correct in the sense that if there is another path $(x_0, z_1, \ldots, z_\ell, x)$ which goes from x_0 to x, then

$$\frac{p(x_0, x_1)p(x_1, x_2) \cdots p(x_k, x)}{p(x_1, x_0)p(x_2, x_1) \cdots p(x, x_k)} = \frac{p(x_0, z_1)p(z_1, z_2) \cdots p(z_\ell, x)}{p(z_1, x_0)p(z_2, z_1) \cdots p(x, z_k)};$$

this, however, immediately follows from the condition (2.6) with the cycle $(x_0, x_1, \ldots, x_k, x, z_\ell, \ldots, z_1, x_0)$. It remains to take some y adjacent to x (as in Figure 2.1) and check that $\pi(x)p(x, y) = \pi(y)p(y, x)$; this is because, by (2.7), $\pi(y) = \pi(x)\frac{p(x,y)}{p(y,x)}$ (note that $(x_0, x_1, \ldots, x_k, x, y)$ is a path from x_0 to y). $\qquad \square$

Now, one of the advantages of reversibility is that, somewhat unexpectedly, it permits us to use physical intuition for analysing the Markov chain. More concretely, it is possible to represent a reversible Markov chain via an electrical network (think of the edges of its transition graph as wires that have some conductance/resistance). In the following, we assume that $p(x, x) = 0$ for all $x \in \Sigma$; this will permit us to avoid loops which do not make much sense in electricity.

Definition 2.3. An *electrical network* is a (non-oriented) graph (V, E) without loops, with positive weights $(c(e), e \in E)$ assigned to its edges. The quantity $c(e)$ is thought of as *conductance* of e, and $r(e) := 1/c(e)$ stands for the *resistance* of the edge e.

An important observation is that we can actually permit the conductances (as well as resistances) to assign values in $[0, +\infty]$ (however, the graph will have to be modified): if $c(e) = 0$ (or $r(e) = \infty$), that means that the edge e is simply removed (no wire at all), and if $c(e) = \infty$ (or $r(e) = 0$) for $e = (x, y)$, that means that the vertices x and y are "glued together" (short-circuited by wires of zero resistance).

The correspondence between reversible Markov chains and electrical networks is described in the following way. Given an electrical network, the transition probabilities of the corresponding Markov chain are then defined by

$$p(x, y) = \frac{c(x, y)}{C(x)}, \quad \text{where } C(x) = \sum_{v:v \sim x} c(x, v). \qquad (2.8)$$

It is then straightforward to verify that it is reversible with the reversible measure C. Conversely, consider a reversible Markov chain with the reversible measure π and with $p(x, x) = 0$ for all x; then define $c(x, y) = \pi(x)p(x, y)$. Also, clearly, the SRW (on any graph) corresponds to the electrical network with all conductances being equal to 1.

We will now take the following route. Instead of proving our way to the desired tools, we will just formulate the necessary facts and explain informally why they should be valid; this is because the author feels that there are already many great sources for learning this theory. First of all, the classical book [39] is an absolute must-read. I can recommend also chapter 9 of [64] for a modern *short* introduction to the subject, and chapters 2, 3, and 9 of [66] for an in-depth treatment.

The central notion of this theory is that of *effective resistance*. The "physical" definition of effective resistance between x and y (denoted by $\mathcal{R}_{\text{eff}}(x, y)$) is simple: attach a 1 volt battery to x and y, measure the (outgoing) current I at x, and define, in accordance with Ohm's law, $\mathcal{R}_{\text{eff}}(x, y) = 1/I$. We can also recall how to calculate resistances of several wires which are put together in the simplest cases of serial and parallel connections; see Figure 2.2.

Next, for a subset $B \subset V$ such that $x \in B \setminus \partial B$, define $\mathcal{R}_{\text{eff}}(x, \partial B)$ to be the effective resistance between x and all vertices of ∂B glued together. Then, $B(x, n)$ being the ball of radius n with respect to the graph distance, define effective resistance to infinity $\mathcal{R}_{\text{eff}}(x, \infty)$ as

$$\mathcal{R}_{\text{eff}}(x, \infty) = \lim_{n \to \infty} \mathcal{R}_{\text{eff}}(x, \partial B(x, n))$$

(the limit exists because that sequence is monotonously increasing, as it is not difficult to see). Intuitively, we just measure the resistance between x and the infinity; see Figure 2.3.

Now, it is possible to prove a remarkable fact that relates the escape

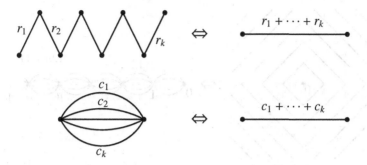

Figure 2.2 Dealing with connections in series and in parallel: in the first case, sum the resistances; in the second case, sum the conductances.

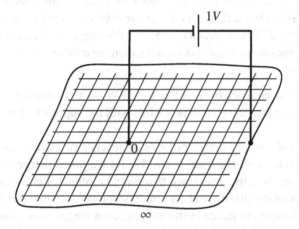

Figure 2.3 On the definition of effective resistance to infinity.

probability from x to the effective resistance:

$$\mathbb{P}_x[\tau_x^+ = \infty] = \frac{1}{C(x)\mathcal{R}_{\text{eff}}(x, \infty)}; \tag{2.9}$$

this implies that the effective resistance to infinity is infinite if and only if the corresponding Markov chain is recurrent.[7]

The following properties of effective resistance are physically evident (but have a non-trivial proof; they are consequences of Rayleigh's Monotonicity Law):

---

[7] Intuitively: if the particle cannot escape to infinity, then the "current" does not flow to infinity, et voilà, the infinite resistance.

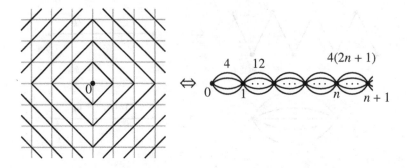

Figure 2.4 Another proof of recurrence in two dimensions.

(i) if we cut certain edges of the network (i.e., set the conductances of
    these edges to 0), then the effective resistance cannot decrease;
(ii) if we "glue" (short-circuit) certain vertices together (i.e., connect them
    with edges of infinite conductance), then the effective resistance cannot
    increase.

A curious consequence of (i) is that the SRW on *any* subgraph of $\mathbb{Z}^2$ is recurrent (just imagine trying to prove this with the approach of the previous
section!).

Let us now obtain another proof of recurrence of the two-dimensional
SRW, this time with electrical networks. As we noted before, we need to
prove that the effective resistance to infinity is infinite. Let us use the previous observation (ii): if we short-circuit some vertices together and prove
that the effective resistance is still infinite, then we are done. Now look at
Figure 2.4: let us glue together the vertices that lie on (graph) distance n
from the origin, for $n = 1, 2, 3, \ldots$. We use the two rules depicted in Figure 2.2: there are $4(2n+1)$ edges from level $n-1$ to level n, so the overall
conductance of the "superedge" between $n-1$ and n is $4(2n+1)$, and so its
resistance is $\frac{1}{4(2n+1)}$. The resistance to infinity of this new graph is therefore

$$\sum_{n=0}^{\infty} \frac{1}{4(2n+1)} = \infty,$$

as required. The harmonic series strikes again!

2.3 Lyapunov functions

The proofs of Sections 2.1 and 2.2 are simple and beautiful. In principle,
this is good, but: not infrequently, such proofs are not very robust, in the

sense that they do not work anymore if the setup is changed even a little bit. Indeed, assume that we modify the transition probabilities of two-dimensional simple random walk in only one site, say, $(1,1)$. For example, let the walk go from $(1,1)$ to $(1,0)$, $(1,2)$, $(0,1)$, $(2,1)$, with probabilities, say, $1/7$, $1/7$, $2/7$, $3/7$, respectively. We keep all other transition probabilities intact. Then, after this apparently innocent change, *both* proofs break down! Indeed, in the classical proof of Section 2.1 the weights of any trajectory that passes through $(1,1)$ would no longer be equal to 4^{-2n}, and so the combinatorics would be hardly manageable (instead of simple formula (2.3), a much more complicated expression will appear). The situation with the proof of Section 2.2 is not very satisfactory as well: random walk is no longer reversible (cf. Exercise 2.8), so the technique of the previous section does not apply at all! It is therefore a good idea to search for a proof which is more robust, i.e., less sensible to small changes of the model's parameters.

In this section, we present a very short introduction to the Lyapunov functions method. Generally speaking, this method consists of finding a function (from the state space of the stochastic process to $\mathbb{R}$) such that the image under this function of the stochastic process is, in some sense, "nice". That is, this new one-dimensional process satisfies some conditions that enable one to obtain results about it and then transfer these results to the original process.

We emphasize that this method is usually "robust", in the sense that the underlying stochastic process need not satisfy simplifying assumptions such as the Markov property, reversibility, or time homogeneity, for instance, and the state space of the process need not be necessarily countable. In particular, this approach works for *non-reversible* Markov chains.

In this section, we follow mainly [71] and [46]. Other sources on the Lyapunov functions method are e.g. [5, 13, 72]; see also [92] for a take on Lyapunov functions from a more applied perspective.

The next result is the main Lyapunov functions tool for proving recurrence of Markov chains.

Theorem 2.4 (Recurrence criterion). *An irreducible Markov chain X_n on a countably infinite state space Σ is recurrent if and only if there exist a function $f : \Sigma \to \mathbb{R}_+$ and a finite nonempty set $A \subset \Sigma$ such that*

$$\mathbb{E}[f(X_{n+1}) - f(X_n) \mid X_n = x] \le 0, \text{ for all } x \in \Sigma \setminus A, \tag{2.10}$$

and $f(x) \to \infty$ as $x \to \infty$.

The quantity in (2.10) can be thought of as the *drift vector* at x with

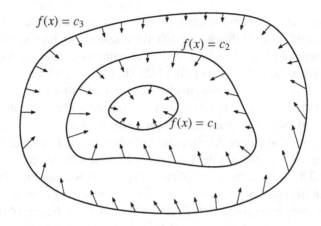

Figure 2.5 The drift vectors point inside the level sets (here, $0 < c_1 < c_2 < c_3$).

respect to the function f. To understand the meaning of Theorem 2.4, recall first that, for a function $f \colon \mathbb{R}^d \mapsto \mathbb{R}$, the level sets are sets of the form $\{x \in \mathbb{R}^d : f(x) = c\}$ for $c \in \mathbb{R}$; in the following heuristics, we think of f as a function of *continuous* argument, just to be able to visualize the things better. If f converges to infinity as $x \to \infty$, then[8] the level sets will look as depicted in Figure 2.5, and Theorem 2.4 says that, to prove the recurrence, it is enough to find a function as shown in Figure 2.5 such that the drift vectors point inside its level sets. In fact, just by observing Figure 2.5 it is easy to believe that the Markov chain should be recurrent, since it has a "tendency" to "go inside".

The term "Lyapunov function" comes from Differential Equations: there, a similar (in spirit) construction is used to prove *stability*[9] of the solutions; see e.g. [62].

Proof of Theorem 2.4   To prove that having a function that satisfies (2.10) is sufficient for the recurrence, let $x \in \Sigma$ be an arbitrary state, and take $X_0 = x$. Let us reason by contradiction, assuming that $\mathbb{P}_x[\tau_A = \infty] > 0$ (which would imply, in particular, that the Markov chain is transient). Set $Y_n = f(X_{n \wedge \tau_A})$ and observe that Y_n is a non-negative supermartingale. Then,

---

[8] The reader may wonder what "$x \to \infty$" might mean on an *arbitrary* countable set, with no particular enumeration fixed. Notice, however, that if a sequence of sites converges to infinity with respect to one enumeration, it will do so with respect to any other one; so, writing "$x \to \infty$" is legitimate in any case.

[9] This can be seen as a deterministic analogue of recurrence.

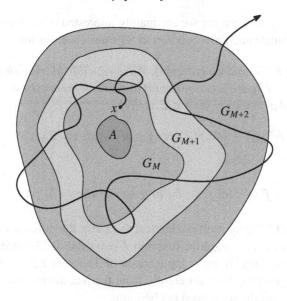

Figure 2.6 On the proof of Theorem 2.4: a transient Markov chain can visit any finite set only finitely many times.

by Theorem 1.5, there exists a random variable Y_∞ such that $Y_n \to Y_\infty$ a.s. and

$$\mathbb{E}_x Y_\infty \leq \mathbb{E}_x Y_0 = f(x), \qquad (2.11)$$

for any $x \in \Sigma$. On the other hand, since $f \to \infty$, it holds that the set $V_M := \{y \in \Sigma : f(y) \leq M\}$ is finite for any $M \in \mathbb{R}_+$; so, our assumption on transience implies that V_M will be visited only finitely many times, meaning that $\lim_{n \to \infty} f(X_n) = +\infty$ a.s. on $\{\tau_A = \infty\}$ (see Figure 2.6). Hence, on $\{\tau_A = \infty\}$, we must have $Y_\infty = \lim_{n \to \infty} Y_n = +\infty$, a.s.. This would contradict (2.11) under the assumption $\mathbb{P}_x[\tau_A = \infty] > 0$, because then $\mathbb{E}_x[Y_\infty] \geq \mathbb{E}_x[Y_\infty \mathbf{1}\{\tau_A = \infty\}] = \infty$. Hence $\mathbb{P}_x[\tau_A = \infty] = 0$ for all $x \in \Sigma$, which means that the Markov chain is recurrent, by Lemma 1.12 (ii).

For the "only if" part (i.e., recurrence implies that there exist f and A as in the preceding instance), see the proof of theorem 2.2.1 of [46]. See also Exercise 2.21. □

As a (very simple) example of application of Theorem 2.4, consider the *one-dimensional* simple random walk $S^{(1)}$, together with the set $A = \{0\}$ and the function $f(x) = |x|$. Then (2.10) holds with equality, which shows that $S^{(1)}$ is recurrent.

Although in this chapter we are mainly interested in the recurrence, let us also formulate and prove a criterion for transience, for future reference:

Theorem 2.5 (Transience criterion). *An irreducible Markov chain X_n on a countable state space Σ is transient if and only if there exist a function $f :$ $\Sigma \to \mathbb{R}_+$ and a nonempty set $A \subset \Sigma$ such that*

$$\mathbb{E}[f(X_{n+1}) - f(X_n) \mid X_n = x] \leq 0, \text{ for all } x \in \Sigma \setminus A, \tag{2.12}$$

and

$$f(y) < \inf_{x \in A} f(x), \text{ for at least one site } y \in \Sigma \setminus A. \tag{2.13}$$

Note that (2.12) by itself is identical to (2.10); the difference is in what we require of the nonnegative function f (we need (2.13) instead of convergence to infinity; in most applications of Theorem 2.5, the function f will converge to 0). Also, differently from the recurrence criterion, in the preceding result the set A need not be finite.

Proof of Theorem 2.5   Assume that $X_0 = y$, where y is from (2.13), and (similarly to the previous proof) define the process $Y_n = f(X_{n \wedge \tau_A})$. Then the relation (2.12) implies that Y_n is a supermartingale. Since Y_n is also non-negative, Theorem 1.5 implies that there is a random variable $Y_\infty \geq 0$ such that $\lim_{n \to \infty} Y_n = Y_\infty$ a.s., and $\mathbb{E}Y_\infty \leq \mathbb{E}Y_0 = f(y)$. Observe that, if the Markov chain eventually hits the set A, then the value of Y_∞ equals the value of f at some random site (namely, X_{τ_A}) belonging to A; formally, we have that, a.s.,

$$Y_\infty \mathbf{1}\{\tau_A < \infty\} = \lim_{n \to \infty} Y_n \mathbf{1}\{\tau_A < \infty\} = f(X_{\tau_A})\mathbf{1}\{\tau_A < \infty\}$$

$$\geq \inf_{x \in A} f(x)\mathbf{1}\{\tau_A < \infty\}.$$

So, we obtain

$$f(y) = \mathbb{E}Y_0 \geq \mathbb{E}Y_\infty \geq \mathbb{E}Y_\infty \mathbf{1}\{\tau_A < \infty\} \geq \mathbb{P}_y[\tau_A < \infty] \inf_{x \in A} f(x),$$

which implies

$$\mathbb{P}_y[\tau_A < \infty] \leq \frac{f(y)}{\inf_{x \in A} f(x)} < 1,$$

proving the transience of the Markov chain X_n, by Lemma 1.12(i).

For the "only if" part, see Exercise 2.15.   □

Let us now think about how to apply Theorem 2.4 to the simple random walk in two dimensions. For this, we need to find a (Lyapunov) function $f: \mathbb{Z}^2 \mapsto \mathbb{R}_+$, such that the "drift with respect to f"

$$\mathbb{E}[f(S_{n+1}) - f(S_n) \mid S_n = x] \tag{2.14}$$

is nonpositive for all but finitely many $x \in \mathbb{Z}^2$, and also such that $f(x) \to \infty$ as $x \to \infty$. The reader must be warned, however, that finding a suitable Lyapunov function is a kind of an art, which usually involves a fair amount of guessing and failed attempts. Still, let us try to understand how it works. In the following, the author will do his best to explain how it *really* works, with all the failed attempts and guessing.

First of all, it is more convenient to think of f as a function of *real* arguments. Now, if there is a general rule of finding a suitable Lyapunov function for processes that live in $\mathbb{R}^d$, then it is the following: consider the level sets of f and think how they should look. In the two-dimensional case, we speak about the level *curves*; of course, we need the function to be sufficiently "good" to ensure that the level curves are really curves in some reasonable sense of this word.

Now, we know that the simple random walk converges to the (two-dimensional) Brownian motion, if suitably rescaled. The Brownian motion is invariant under rotations, so it seems reasonable to search for a function that only depends on the Euclidean norm of the argument, $f(x) = g(\|x\|)$ for some *increasing* function $g: \mathbb{R} \mapsto \mathbb{R}_+$. Even if we did not know about the Brownian motion, it would still be reasonable to make this assumption because, well, why not? It is easier to make calculations when there is some symmetry. Notice that, in this case, the level curves of f are just circles centred at the origin.[10]

So, let us begin by looking at the level curves of a very simple function $f(x) = \|x\|$, and see what happens to the drift (2.14). Actually, let us just look at Figure 2.7; the level curves shown are $\{\|x\| = k - 1\}$, $\{\|x\| = k\}$, $\{\|x\| = k + 1\}$ on the right, and $\{\|x\| = \sqrt{j^2 + (j-1)^2}\}$, $\{\|x\| = j\sqrt{2}\}$, $\{\|x\| = 2\sqrt{2}j - \sqrt{j^2 + (j-1)^2}\}$ on the left.[11] It is quite clear then that the drift with respect to $f(x) = \|x\|$ is strictly positive in both cases. Indeed, one sees that, in the first case, the jumps to the left and to the right "compensate" each other, while the jumps up and down both slightly increase the norm. In the second case, jumps up and to the left change the norm by a larger amount

---

[10] There are many examples where they are not circles/spheres; let us mention e.g. section 4.3 of [71], which is based on [74].

[11] Observe that, similarly to the previous case, these level curves have form $\{\|x\| = a - b\}$, $\{\|x\| = a\}$, $\{\|x\| = a + b\}$ with $a = j\sqrt{2}, b = j\sqrt{2} - \sqrt{j^2 + (j-1)^2}$.

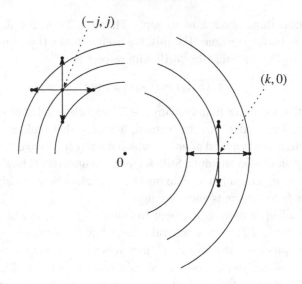

Figure 2.7 The drift with respect to $f(x) = \|x\|$ is positive.

than the jumps down and to the right. In fact, it is possible to prove that the drift is positive for all $x \in \mathbb{Z}^2$, but the preceding examples show that, for proving the recurrence, the function $f(x) = \|x\|$ will not work anyway.

Now, think e.g. about the "diagonal case": if we move the third level curve a little bit outside, then the drift with respect to the function would become nonpositive; look at Figure 2.8. It seems to be clear that, to produce such level curves, the function g should have a sublinear growth. Recall that we are "guessing" the form that g may have, so such nonrigorous reasoning is perfectly acceptable; we just need to find a function that works, and the way how we arrived to it is totally unimportant from the formal point of view. A natural first candidate would be then $g(s) = s^\alpha$, where $\alpha \in (0, 1)$. So, let us try it! Let $x \in \mathbb{Z}^2$ be such that $\|x\|$ is large, and let e be a unit vector (actually, it is $\pm e_1$ or $\pm e_2$). Write (being $y \cdot z$ the scalar product of $y, z \in \mathbb{Z}^2$)

$$\|x + e\|^\alpha - \|x\|^\alpha = \|x\|^\alpha \left(\left(\frac{\|x + e\|}{\|x\|} \right)^\alpha - 1 \right)$$
$$= \|x\|^\alpha \left(\left(\frac{(x + e) \cdot (x + e)}{\|x\|^2} \right)^{\alpha/2} - 1 \right)$$
$$= \|x\|^\alpha \left(\left(\frac{\|x\|^2 + 2x \cdot e + 1}{\|x\|^2} \right)^{\alpha/2} - 1 \right)$$

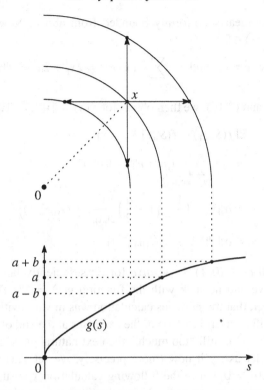

Figure 2.8 What should the function g look like? (For x on the diagonal, we have $a = g(\|x\|)$, $a \pm b = g(\|x \pm e_1\|)$; note that $\|x + e_1\| - \|x\| > \|x\| - \|x - e_1\|$.)

$$= \|x\|^\alpha\Big(\Big(1 + \frac{2x \cdot e + 1}{\|x\|^2}\Big)^{\alpha/2} - 1\Big).$$

Note that $|x \cdot e| \le \|x\|$, so the term $\frac{2x \cdot e + 1}{\|x\|^2}$ should be small (at most $O(\|x\|^{-1})$); let us also recall the Taylor expansion $(1+y)^{\alpha/2} = 1 + \frac{\alpha}{2}y - \frac{\alpha}{4}(1 - \frac{\alpha}{2})y^2 + O(y^3)$. Using that, we continue the preceding calculation:

$$\|x + e\|^\alpha - \|x\|^\alpha$$

$$= \|x\|^\alpha\Big(\frac{\alpha x \cdot e}{\|x\|^2} + \frac{\alpha}{2\|x\|^2} - \frac{\alpha}{4}\Big(1 - \frac{\alpha}{2}\Big)\frac{(2x \cdot e + 1)^2}{\|x\|^4} + O(\|x\|^{-3})\Big)$$

$$= \alpha\|x\|^{\alpha-2}\Big(x \cdot e + \frac{1}{2} - \Big(1 - \frac{\alpha}{2}\Big)\frac{(x \cdot e)^2}{\|x\|^2} + O(\|x\|^{-1})\Big). \tag{2.15}$$

Observe that in the preceding display the $O(\cdot)$'s actually depend also on the direction of x (that is, the unit vector $x/\|x\|$), but this is not a problem since they are clearly uniformly bounded from above. Now notice that, with $x = (x_1, x_2) \in \mathbb{Z}^2$,

$$\sum_{e \in \{\pm e_1, \pm e_2\}} x \cdot e = 0, \quad \text{and} \quad \sum_{e \in \{\pm e_1, \pm e_2\}} (x \cdot e)^2 = 2x_1^2 + 2x_2^2 = 2\|x\|^2. \quad (2.16)$$

Using (2.15) and (2.16), we then obtain for $f(x) = \|x\|^\alpha$, as $\|x\| \to \infty$,

$$\mathbb{E}[f(S_{n+1}) - f(S_n) \mid S_n = x]$$

$$= \frac{1}{4} \sum_{e \in \{\pm e_1, \pm e_2\}} (\|x + e\|^\alpha - \|x\|^\alpha)$$

$$= \alpha\|x\|^{\alpha-2}\Big(\frac{1}{2} - \Big(1 - \frac{\alpha}{2}\Big)\frac{\|x\|^2}{2\|x\|^2} + O(\|x\|^{-1})\Big)$$

$$= \alpha\|x\|^{\alpha-2}\Big(\frac{\alpha}{4} + O(\|x\|^{-1})\Big), \quad (2.17)$$

which, for all $\alpha \in (0, 1)$, is *positive* for all sufficiently large x. So, unfortunately, we had no luck with the function $g(s) = s^\alpha$. That does not mean, however, that the previous calculation was in vain; with some small changes, it will be useful for one of the exercises at the end of this chapter.

Since $g(s) = s^\alpha$ is still "too much", the next natural guess is $g(s) = \ln s$ then.[12] Well, let us try it now (more precisely, we set $f(x) = \ln\|x\|$ for $x \neq 0$ and $f(0) = 0$, but in the following calculation x is supposed to be far from the origin in any case). Using the Taylor expansion $\ln(1 + y) = y - \frac{1}{2}y^2 + O(y^3)$, we write

$$\ln\|x + e\| - \ln\|x\| = \ln\frac{(x + e) \cdot (x + e)}{\|x\|^2}$$

$$= \ln\Big(1 + \frac{2x \cdot e + 1}{\|x\|^2}\Big)$$

$$= \frac{2x \cdot e}{\|x\|^2} + \frac{1}{\|x\|^2} - \frac{2(x \cdot e)^2}{\|x\|^4} + O(\|x\|^{-3}), \quad (2.18)$$

so, using (2.16) again, we obtain (as $x \to \infty$)

$$\mathbb{E}[f(S_{n+1}) - f(S_n) \mid S_n = x] = \frac{1}{4} \sum_{e \in \{\pm e_1, \pm e_2\}} (\ln\|x + e\| - \ln\|x\|)$$

$$= \frac{1}{\|x\|^2} - \frac{1}{4} \times \frac{4\|x\|^2}{\|x\|^4} + O(\|x\|^{-3})$$

[12] The reader may have recalled that $\ln\|x\|$ is a harmonic function in two dimensions, so $\ln\|B_t\|$ is a (local) martingale, where B is a two-dimensional standard Brownian motion. But, lattice effects may introduce some corrections. . .

$$= O(\|x\|^{-3}),$$

which gives us absolutely nothing. Apparently, we need more terms in the Taylor expansion, so let us do the work: with $\ln(1 + y) = y - \frac{1}{2}y^2 + \frac{1}{3}y^3 - \frac{1}{4}y^4 + O(y^5)$, we have[13]

$$\ln \|x + e\| - \ln \|x\| = \ln \left(1 + \frac{2x \cdot e + 1}{\|x\|^2}\right)$$

$$= \frac{2x \cdot e}{\|x\|^2} + \frac{1}{\|x\|^2} - \frac{2(x \cdot e)^2}{\|x\|^4} - \frac{2x \cdot e}{\|x\|^4} - \frac{1}{2\|x\|^4}$$

$$+ \frac{8(x \cdot e)^3}{3\|x\|^6} + \frac{4(x \cdot e)^2}{\|x\|^6} - \frac{4(x \cdot e)^4}{\|x\|^8} + O(\|x\|^{-5}).$$

Then, using (2.16) together with the fact that

$$\sum_{e \in \{\pm e_1, \pm e_2\}} (x \cdot e)^3 = 0, \quad \text{and} \quad \sum_{e \in \{\pm e_1, \pm e_2\}} (x \cdot e)^4 = 2(x_1^4 + x_2^4),$$

we obtain

$$\mathbb{E}[f(S_{n+1}) - f(S_n) \mid S_n = x]$$

$$= \frac{1}{\|x\|^2} - \frac{1}{\|x\|^2} - \frac{1}{2\|x\|^4} + \frac{2\|x\|^2}{\|x\|^6} - \frac{2(x_1^4 + x_2^4)}{\|x\|^8} + O(\|x\|^{-5})$$

$$= \|x\|^{-4}\left(\frac{3}{2} - \frac{2(x_1^4 + x_2^4)}{\|x\|^4} + O(\|x\|^{-1})\right). \tag{2.19}$$

We want the right-hand side of (2.19) to be nonpositive for all x large enough, and it is indeed so if x is on the axes or close enough to them (for $x = (a, 0)$ or $(0, a)$, the expression in the parentheses becomes $\frac{3}{2} - 2 + O(\|x\|^{-1}) < 0$ for all large enough x). Unfortunately, when we check it for the "diagonal" sites (i.e., $x = (\pm a, \pm a)$, so that $\frac{2(x_1^4 + x_2^4)}{\|x\|^4} = \frac{2(a^4 + a^4)}{4a^4} = 1$), we obtain that the expression in the parentheses is $\frac{3}{2} - 1 + O(\|x\|^{-1})$, which is *strictly positive* for all large enough x.

So, this time we were quite close, but still missed the target. A next natural candidate would be a function that grows even slower than the logarithm; so, let us try the function $f(x) = \ln^\alpha \|x\|$ with $\alpha \in (0, 1)$. Hoping for the best, we write (using $(1 + y)^\alpha = 1 + \alpha y - \frac{\alpha(1-\alpha)}{2}y^2 + O(y^3)$ in the last passage)

$$\ln^\alpha \|x + e\| - \ln^\alpha \|x\|$$

$$= \ln^\alpha \|x\|\left(\frac{\ln^\alpha \|x + e\|}{\ln^\alpha \|x\|} - 1\right)$$

---

[13] The reader is invited to check that only one extra term is not enough.

$$= \ln^\alpha \|x\| \left(\left(\frac{\ln\left(\|x\|^2\left(1 + \frac{2x \cdot e + 1}{\|x\|^2}\right)\right)}{\ln \|x\|^2} \right)^\alpha - 1 \right)$$

$$= \ln^\alpha \|x\| \left(\left(1 + (\ln \|x\|^2)^{-1} \ln\left(1 + \frac{2x \cdot e + 1}{\|x\|^2}\right)\right)^\alpha - 1 \right)$$

(by (2.18))

$$= \ln^\alpha \|x\| \left(\left(1 + (\ln \|x\|^2)^{-1} \left(\frac{2x \cdot e}{\|x\|^2} + \frac{1}{\|x\|^2}\right. \right.\right.$$
$$\left.\left.\left. - \frac{2(x \cdot e)^2}{\|x\|^4} + O(\|x\|^{-3})\right)\right)^\alpha - 1 \right)$$

$$= \ln^\alpha \|x\| \left(\alpha(\ln \|x\|^2)^{-1} \left(\frac{2x \cdot e}{\|x\|^2} + \frac{1}{\|x\|^2} - \frac{2(x \cdot e)^2}{\|x\|^4} + O(\|x\|^{-3})\right) \right.$$
$$\left. - \frac{\alpha(1 - \alpha)}{2} (\ln \|x\|^2)^{-2} \frac{4(x \cdot e)^2}{\|x\|^4} + O(\|x\|^{-3}(\ln \|x\|)^{-2}) \right).$$

Then, using (2.16), we obtain

$$\mathbb{E}[f(S_{n+1}) - f(S_n) \mid S_n = x]$$

$$= \frac{\alpha}{2} \ln^{\alpha-1} \|x\| \left(\frac{1}{\|x\|^2} - \frac{\|x\|^2}{\|x\|^4} + O(\|x\|^{-3}) \right.$$

$$\left. - \frac{(1 - \alpha)}{2} (\ln \|x\|^2)^{-1} \frac{2\|x\|^2}{\|x\|^4} + O((\|x\| \ln \|x\|)^{-2}) \right)$$

$$= -\frac{\alpha}{2\|x\|^2 \ln^{2-\alpha} \|x\|} \left(\frac{(1 - \alpha)}{2} + O(\|x\|^{-1} \ln \|x\|) \right),$$

which is[14] negative for all sufficiently large x. Thus Theorem 2.4 shows that SRW on $\mathbb{Z}^2$ is recurrent, proving Pólya's theorem (Theorem 1.1) in the two-dimensional case.

Now it is time to explain why the author likes this method of proving recurrence (and many other things) of countable Markov chains. First, observe that the preceding proof does not use any trajectory-counting arguments (as in Section 2.1) or reversibility (as in Section 2.2), recall the example in the beginning of this section. Moreover, consider *any* Markov chain X_n on the two-dimensional integer lattice with asymptotically zero drift, and let us abbreviate $D_x = X_1 - x$. Analogously to the preceding proof, we can obtain (still using $f(x) = \ln^\alpha \|x\|$ with $\alpha \in (0, 1)$)

$$\mathbb{E}[f(X_{n+1}) - f(X_n) \mid X_n = x]$$

[14] Finally!

$$= -\frac{\alpha}{\|x\|^2 \ln^{2-\alpha} \|x\|} \Big(-\ln\|x\|^2 \mathbb{E}_x x \cdot D_x - \ln\|x\|^2 \mathbb{E}_x \|D_x\|^2$$
$$+ \ln\|x\|^2 \frac{2\mathbb{E}_x (x \cdot D_x)^2}{\|x\|^2} + 2(1-\alpha)\frac{\mathbb{E}_x (x \cdot D_x)^2}{\|x\|^2} + O((\ln\|x\|)^{-1}) \Big).$$

Now, if we can prove that the expression in the parentheses is positive for all large enough x, then this would imply the recurrence. It seems to be clear that it will be the case if the transitions probabilities at x are sufficiently close to those of the simple random walk (and the difference converges to 0 sufficiently fast as $x \to \infty$). This is what we meant when saying that the method of Lyapunov functions is robust: if it works for a particular model (the simple random walk in two dimensions, in our case), then one may expect that the same (or almost the same) Lyapunov function will also work for "close" models. See also Exercise 2.20 for some further ideas.

Also, besides proving recurrence/transience, Lyapunov functions may be useful for doing many other things; see e.g. [24, 68, 70] as well as [71].

2.4 Exercises

Combinatorial proofs (Section 2.1)

Exercise 2.1. Understand the original proof of Pólya [76]. (Warning: it uses generating functions, and it is in German.)

Exercise 2.2. Let $p_{2n}^{(d)}$ be the probability that d-dimensional SRW finds itself at its starting position after $2n$ steps. In the end of Section 2.1, we have seen that $p_{2n}^{(2)} = (p_{2n}^{(1)})^2$, because it is possible to decouple the components of a two-dimensional SRW. Can this decoupling be done for at least *some* $d \geq 3$ as well (so that, in particular, $p_{2n}(0,0) = (p_{2n}^{(1)}(0,0))^d$ would hold)?

Exercise 2.3. Find a direct (combinatorial) proof of the recurrence of simple random walk on some other regular lattice (e.g., triangular, hexagonal, etc.) in two dimensions.

Exercise 2.4. If S_n is the two-dimensional SRW, prove that $S_n \cdot (e_1 + e_2)$ and $S_n \cdot (e_1 - e_2)$ are *independent* one-dimensional SRWs.

Exercise 2.5. Using the result of the previous exercise, can one derive a version of *local* Central Limit Theorem for the two-dimensional SRW from the de Moivre–Laplace theorem?

Exercise 2.6. Find a "direct" proof of recurrence of two-dimensional SRW (i.e., not using the fact that, for any Markov chain, the total number of visits to a site has geometric distribution, which permitted us to relate the actual number of visits to the expected number of visits) by showing that there exists $\delta > 0$ such that for any n_0

$$\mathbb{P}[\text{there is } k \geq n_0 \text{ such that } S_k = 0] \geq \delta.$$

Suggestion: use (a particular case of) the Paley–Zygmund inequality: if Z is a nonnegative random variable with finite second moment, then $\mathbb{P}[Z > 0] \geq (\mathbb{E}Z)^2/\mathbb{E}Z^2$. Then, given n_0, find $n_1 > n_0$ such that the Paley–Zygmund inequality works well when applied on the random variable $\sum_{k=n_0}^{n_1} \mathbf{1}\{S_k = 0\}$.

Reversibility and electrical networks (Section 2.2)

Exercise 2.7. Consider a nearest-neighbour random walk on $\mathbb{Z}$ with transition probabilities $p_x = p(x, x - 1)$, $q_x = 1 - p_x = p(x, x + 1)$. Show that it is reversible and write its reversible measure.

Exercise 2.8. Show that the random walk described in the beginning of Section 2.3 (the one where we changed the transition probabilities at the site $(1, 1)$) is not reversible.

Exercise 2.9. Consider a reversible Markov chain on a finite state space Σ with reversible measure π. Let us think of it as a linear operator P acting on functions $f : \Sigma \to \mathbb{R}$ in the following way: $Pf(x) = \sum_{y\in\Sigma} p(x, y)f(y)$. Prove that all the eigenvalues of P are real and belong to $[-1, 1]$.

Exercise 2.10. Consider an irreducible reversible Markov chain on a countable state space Σ with the reversible measure π, and let $0 \in \Sigma$ be a fixed state. Assume that $\Sigma = \bigcup_{k=0}^{\infty} \Sigma_k$, where $\Sigma_0 = \{0\}$ and Σ_ks are disjoint. Further, assume that for all k

$$\tilde{\pi}_k := \sum_{x\in\Sigma_k} \pi(x) < \infty,$$

and $\mathbb{P}_0[\tau_{R_k^c} < \infty] = 1$, where $R_k := \bigcup_{j=1}^{k} \Sigma_k$. Let us define another Markov chain $\tilde{X}$ on the state space $\mathbb{Z}_+$ with transition probabilities

$$\tilde{p}(k, \ell) = \frac{1}{\tilde{\pi}_k} \sum_{\substack{x\in\Sigma_k \\ y\in\Sigma_\ell}} \pi(x)p(x, y).$$

Intuitively, this new Markov chain works in the following way: when the original process is in Σ_k, we resample its location according to the distribution $\pi(\cdot)/\tilde{\pi}_k$; the process $\tilde{X}$ then just "observes" in which of the Σs is this "modified" walker.

Prove that the Markov chain $\tilde{X}$ is irreducible on $\mathbb{Z}_+$ and reversible with the reversible measure $\tilde{\pi}$.

Exercise 2.11. In the setting of the previous exercise, theorem 6.10 of [65] implies that if the new Markov chain $\tilde{X}$ is recurrent, then so is the original Markov chain. Apply this result to obtain yet another proof of recurrence of two-dimensional SRW.

Exercise 2.12. Show that, for any $a, b, c, d > 0$

$$\frac{(a+c)(b+d)}{a+b+c+d} \geq \frac{ab}{a+b} + \frac{cd}{c+d} \tag{2.20}$$

(the reader has probably figured out that the idea is to find an "electrical proof" of this inequality).

Exercise 2.13. There are also ways to prove transience using electrical networks; review theorem 2.11 of [66] and, using it, prove that SRW in three dimensions (and, consequently, in any dimension $d \geq 3$) is transient.

Exercise 2.14. Fix $\varepsilon \in (0, 1]$, and consider the set $\{x \in \mathbb{Z}^3 : |x \cdot e_3| \leq (|x \cdot e_1| + |x \cdot e_2|)^\varepsilon\}$ with the graph structure inherited from $\mathbb{Z}^3$. Prove that the SRW on this set is transient. (It can be easily seen that the size of the ball of radius r centred in the origin with respect to the graph distance is $O(r^{2+\varepsilon})$; so the preceding can be informally seen as "SRW in dimension $2 + \varepsilon$".)

Lyapunov functions (Section 2.3)

Exercise 2.15. Prove the "only if" part of the transience criterion (Theorem 2.5).

Exercise 2.16. Find a Lyapunov-function proof of transience of a nearest-neighbour random walk on $\mathbb{Z}$ with constant drift.

Exercise 2.17. Now, consider a Markov chain (X_n) on $\mathbb{Z}_+$ such that

- there exists $K > 0$ such that $|X_{n+1} - X_n| \leq K$ a.s. (i.e., the jumps are uniformly bounded);
- there exists $\varepsilon > 0$ such that $\mathbb{E}_x X_1 \geq x + \varepsilon$ for all x (i.e., the drift is uniformly positive).

Prove that it is transient.

Exercise 2.18. Using Theorem 2.5, prove that simple random walk in dimensions $d \geq 3$ is transient. Hint: use $f(x) = \|x\|^{-\alpha}$ for some $\alpha > 0$.

Exercise 2.19. Show that $f(x) = \ln \ln \|x\|$ (suitably redefined at the origin and at the sites which are at distance at most e from it) would also work for proving the recurrence of two-dimensional simple random walk.

Exercise 2.20. Using Lyapunov functions, prove the recurrence of a two-dimensional spatially homogeneous zero-mean random walk with bounded jumps.

Exercise 2.21. Understand the proof of the "only if" part of the recurrence criterion (Theorem 2.4) — see the proof of theorem 2.2.1 of [46]. Can you find[15] a simpler proof?

Exercise 2.22. Is it always possible to find a function which is a *martingale* outside the origin (i.e., in Theorem 2.4, A is a singleton and (2.4) holds with equality) for a recurrent Markov chain? Also, prove that this is possible for a recurrent Markov chain on $\mathbb{Z}_+$ with nearest-neighbour jumps by writing down such a function explicitly.

Exercise 2.23. The following result (also known as *Foster's criterion* or *Foster–Lyapunov theorem*) provides a criterion for the positive recurrence of an irreducible Markov chain:

An irreducible Markov chain X_n on a countable state space Σ is positive recurrent if and only if there exist a positive function $f : \Sigma \to \mathbb{R}_+$, a finite nonempty set $A \subset \Sigma$, and $\varepsilon > 0$ such that

$$\mathbb{E}[f(X_{n+1}) - f(X_n) \mid X_n = x] \leq -\varepsilon, \text{ for all } x \in \Sigma \setminus A, \qquad (2.21)$$

$$\mathbb{E}[f(X_{n+1}) \mid X_n = x] < \infty, \text{ for all } x \in A. \qquad (2.22)$$

(i) Prove the "only if" part.
(ii) Understand the proof of the "if" part (see e.g. theorems 2.6.2 and 2.6.4 of [71]).

Exercise 2.24. Consider the function $f : \mathbb{N} \to \mathbb{N}$, defined in the following way:

$$f(n) = \begin{cases} \frac{n}{2}, & \text{if } n \text{ is even,} \\ \frac{3n+1}{2}, & \text{if } n \text{ is odd.} \end{cases}$$

The (yet unproven, and very difficult) Collatz conjecture asserts that, for any $n \in \mathbb{N}$, the sequence $f(n), f(f(n)), f(f(f(n))), \ldots$ will eventually reach 1.

(i) Do *not* try to prove the Collatz conjecture.

---

[15] If you find it, please, let me know.

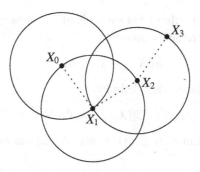

Figure 2.9 A random walk on $\mathbb{R}^2$.

(ii) Consider, however, the stochastic process $(Y_n, n \geq 1)$ living on $(0, +\infty)$, defined as follows:

$$Y_{n+1} = \begin{cases} \frac{Y_n}{2}, & \text{with probability } \frac{1}{2}, \\ \frac{3Y_n+1}{2}, & \text{with probability } \frac{1}{2}. \end{cases}$$

Prove that this process is "positive recurrent", in the sense that $\mathbb{E}(\min\{k : Y_k \leq 1\} \mid Y_0 = y) < \infty$ for any $y \geq 1$.

(Since, informally, "half of all natural numbers are even", the recurrence of the process in (ii) may be considered as an "empirical evidence" in favour of the Collatz conjecture. See [58] and references therein for much more discussion on stochastic models "related" to the Collatz conjecture.)

Exercise 2.25. Let $(X_n, n \geq 0)$ be a discrete-time Markov process on the (continuous) state space $\mathbb{R}^2$, defined in the following way. Let $X_0 = 0$ and, given that $X_n = x \in \mathbb{R}^2$, the next state X_{n+1} is uniformly distributed on $\partial B(x, 1)$ (see Figure 2.9). Prove that this process is *Harris recurrent*, in the sense that it visits any fixed neighbourhood of the origin (and, consequently, any fixed neighbourhood of any other point) infinitely many times a.s..

Exercise 2.26. For the d-dimensional simple random walk, show that the first and the second moments of $\Delta_x := \|S_1\| - \|x\|$ under $\mathbb{E}_x$ are given by

$$\mathbb{E}_x \Delta_x = \frac{d-1}{2d\|x\|} + O(\|x\|^{-2}), \tag{2.23}$$

$$\mathbb{E}_x \Delta_x^2 = \frac{1}{d} + O(\|x\|^{-1}). \tag{2.24}$$

Exercise 2.27. Suppose now that $(X_n, n \geq 0)$ is a time homogeneous

Markov chain on an unbounded subset Σ of $\mathbb{R}_+$. Assume that X_n has uniformly bounded increments, so that

$$\mathbb{P}[|X_{n+1} - X_n| \le b] = 1 \tag{2.25}$$

for some $b \in \mathbb{R}_+$. For $k = 1, 2$ define

$$\mu_k(x) := \mathbb{E}[(X_{n+1} - X_n)^k \mid X_n = x].$$

The first moment function, $\mu_1(x)$, is also called the one-step *mean drift* of X_n at x.

Lamperti [59, 60, 61] investigated the extent to which the asymptotic behaviour of such a process is determined by $\mu_{1,2}(\cdot)$, in a typical situation when $\mu_1(x) = O(x^{-1})$ and $\mu_2(x) = O(1)$. The following three statements are particular cases of Lamperti's fundamental results on recurrence classification:

(i) If $2x\mu_1(x) + \mu_2(x) < -\varepsilon$ for some $\varepsilon > 0$ and all large enough x, then X_n is positive recurrent;

(ii) If $2x\mu_1(x) - \mu_2(x) < -\varepsilon$ for some $\varepsilon > 0$ and all large enough x, then X_n is recurrent;

(iii) If $2x\mu_1(x) - \mu_2(x) > \varepsilon$ for some $\varepsilon > 0$ and all large enough x, then X_n is transient.

Prove (i), (ii), and (iii).

Exercise 2.28. Let $d \ge 3$. Show that for any $\varepsilon > 0$ there exists $C = C(d, \varepsilon)$ such that $\|S_{n \wedge \tau}\|^{-(d-2)-\varepsilon}$ is a submartingale and $\|S_{n \wedge \tau}\|^{-(d-2)+\varepsilon}$ is a supermartingale, where $\tau = \tau^+_{\mathsf{B}(C)}$. What happens in the case $\varepsilon = 0$?

3

Some potential theory for simple random walks

Disclaimer: this chapter is by no means a systematic treatment of the subject, not even remotely so. If the reader is looking for one, the author can recommend e.g. [38, 86] or chapters 4 and 6 of [63]. Here we rather adopt a "customer's point of view": we only recall a few general notions and tools that permit us to obtain estimates on what we are concerned about in this book – probabilities related to simple random walks. We also do not try to discuss the reason why it is called "potential theory" and what exactly are its relations to the classical theory of harmonic functions – this would take quite some time, and the author has to confess that he does not understand it quite well anyway.

We are going to consider the transient and the recurrent cases separately; it is true that in this book we are mainly concerned with the latter one, but it is still more convenient to begin with the transient case (i.e., SRW in dimensions $d \geq 3$), which is somehow conceptually simpler.[1] Let us begin, though, with the following result, which is dimension independent:

Proposition 3.1. *Let $h \colon \mathbb{Z}^d \to \mathbb{R}$ be a harmonic function, i.e.,*

$$h(x) = \frac{1}{2d} \sum_{y \sim x} h(y) \qquad \text{for all } x \in \mathbb{Z}^d. \tag{3.1}$$

Assume also that it is bounded, i.e., there exists $K > 0$ such that $|h(x)| \leq K$ for all $x \in \mathbb{Z}^d$. Then h is constant.

Proof Let us reason by contradiction: assume that a function h as in this proposition is *not* constant. Then it must be nonconstant on both the set of even[2] sites and the set of odd sites. Indeed, its value on an even site is the average of its values on the neighbouring odd sites and vice versa; so, if it is equal to a constant on one of these two sets, it will have to be

---

[1] Also, we'll need some of these transient-case results in Chapter 6 when discussing the "classical" random interlacement model in dimensions $d \geq 3$.

[2] Recall that even (odd) sites are sites with even (odd) sum of their coordinates.

the same constant on the other one. Now, we can find two sites x and y of the same parity such that $x - y = \pm 2e_k$ for some $k \in \{1, \ldots, d\}$ and $h(x) \neq h(y)$. Note that, if h is harmonic in the sense of (3.1), then so are its translations/rotations/reflections. Therefore, without restricting generality, we can assume that $h(e_1) \neq h(-e_1)$.

For $x = (x_1, x_2, \ldots, x_d) \in \mathbb{Z}^d$, let us denote $\bar{x} = (-x_1, x_2, \ldots, x_d)$ the "mirrored" site with respect to the hyperplane orthogonal to the first coordinate axis. Let (S_n) be the SRW starting at e_1, and $(\bar{S}_n)$ be the corresponding "mirrored" SRW (which, clearly, starts at $(-e_1)$). Define the stopping time

$$\sigma = \min\{k : S_k \cdot e_1 = 0\} = \min\{k : S_k = \bar{S}_k\}$$

to be the moment when S meets $\bar{S}$. Note that $\sigma < \infty$ a.s., due to the recurrence of one-dimensional SRW.

Now, the harmonicity of h implies that $h(S_n)$ is a martingale: indeed, $\mathbb{E}(h(S_{n+1}) \mid S_n = x) = (2d)^{-1} \sum_{y \sim x} h(y) = h(x)$ by (3.1); clearly, the same argument also implies that $h(\bar{S}_n)$ is a martingale. Since h is bounded, Corollary 1.7 implies that

$$h(e_1) = \mathbb{E}h(S_\sigma) = \mathbb{E}h(\bar{S}_\sigma) = h(-e_1),$$

which is the desired contradiction (recall that we just assumed that $h(x) \neq h(y)$).   □

3.1 Transient case

First, let us go to dimensions $d \geq 3$, where the SRW is transient. We specifically concentrate on the simple random walk case, although a similar theory can be developed for random walks with arbitrary jump distribution, or even general transient *reversible* Markov chains. We need first to recall some basic definitions related to simple random walks in higher dimensions.

The three main objects that we need are Green's function, the capacity, and the harmonic measure. Specifically, it holds the following:

- Green's function G is harmonic outside the origin, and so the process $G(S_{n \wedge \tau_0})$ is a martingale. This gives us a convenient tool for calculating certain exit probabilities via the optional stopping theorem.[3]
- Informally speaking, the capacity of a set measures how big this set is

---

[3] Stated as Theorem 1.6 in this book; in fact, Corollary 1.7 will usually be enough for our needs.

from the point of view of the simple random walk. This permits us to obtain some refined bounds on probabilities of hitting sets.

• The harmonic measure is a probability distribution that lives on the boundary of a set, and is the "*conditional* entrance measure from infinity". When we are concerned with entrance measures starting from some fixed site, sometimes it is possible to argue that this entrance measure is not much different from the harmonic measure, thus allowing us to have some control on where the random walk enters that set.

Let us now elaborate.

Green's function

For $d \geq 3$, Green's function $G : (\mathbb{Z}^d)^2 \to \mathbb{R}_+$ is defined in the following way:

$$G(x,y) = \mathbb{E}_x\Big(\sum_{k=0}^{\infty} \mathbf{1}\{S_k = y\}\Big) = \sum_{k=0}^{\infty} \mathbb{P}_x[S_k = y]. \qquad (3.2)$$

That is, $G(x,y)$ is equal to the mean number of visits to y starting from x. It is important to note that in the case $x = y$ we do count this as one "initial" visit (so, in particular, $G(x,x) > 1$ in all dimensions). By symmetry it holds that $G(x,y) = G(y,x) = G(0, y-x)$; thus, we can abbreviate $G(y) := G(0,y)$ so that $G(x,y) = G(x-y) = G(y-x)$. Now, a very important property of $G(\cdot)$ is that it is harmonic outside the origin, i.e.,

$$G(x) = \frac{1}{2d} \sum_{y \sim x} G(y) \quad \text{for all } x \in \mathbb{Z}^d \setminus \{0\}. \qquad (3.3)$$

Since, as observed earlier, $G(x)$ is the mean number of visits to the origin starting from x, one readily obtains the preceding from the total expectation formula, with only a little bit of thinking in the case when x is a neighbour of the origin. An immediate consequence of (3.3) is the following.

Proposition 3.2. *The process $G(S_{n \wedge \tau_0})$ is a martingale.*

Now, how should $G(x)$ behave as $x \to \infty$? It is (almost) clear that it converges to 0 by transience, but how fast? It is not difficult to see[4] that $G(x)$ must be of order $\|x\|^{-(d-2)}$, due to the following heuristic argument. Fix $x \in \mathbb{Z}^d$, $x \neq 0$. First, as is well known (think e.g. of the Central Limit Theorem), the simple random walk is *diffusive*, i.e., it needs time of order $\|x\|^2$ to be able to deviate from its initial position by distance $\|x\|$ (which is a necessary condition if it wants to go to x). Then, at time $m > \|x\|^2$ the walk

---

[4] "To see" does not mean "to prove".

can be anywhere[5] in a ball of radius roughly $m^{1/2}$, which has volume of order $m^{d/2}$. So, the chance that the walk is in x should be[6] of order $m^{-d/2}$; therefore, Green's function's value in x is roughly

$$\sum_{m=\|x\|^2}^{\infty} m^{-d/2} \asymp (\|x\|^2)^{-d/2+1} = \|x\|^{-(d-2)}.$$

Note also that

$$G(x) = \mathbb{P}_x[\tau_0 < \infty]G(0); \tag{3.4}$$

indeed, starting from x, the mean number of visits to 0 is zero given that $\tau_0 = \infty$ and is $G(0)$ given that $\tau_0 < \infty$, so the preceding again comes out from the total expectation formula. This implies that the probability of ever visiting y starting from x (which is the same as the probability of ever visiting 0 starting from $x - y$) is also of order $\|x - y\|^{-(d-2)}$.

Now, since the SRW is "roughly spherically symmetric", it is reasonable to expect that Green's function should be asymptotically "well behaved" and, in particular, depend (almost) only on $\|x\|$ as $x \to \infty$. In fact, it is possible to obtain that

$$G(x) = \frac{\gamma_d}{\|x\|^{d-2}} + O(\|x\|^{-d}), \tag{3.5}$$

with $\gamma_d = \frac{\Gamma(d/2)d}{\pi^{d/2}(d-2)}$; see theorem 4.3.1 of [63]. We prefer not to include the complete proof here (that is, we will take this fact for granted), but see Exercises 3.3 through 3.5.

We are now able to obtain a straightforward (*and* useful) estimate for the probability that the simple random walk escapes an annulus through its outer boundary:

Lemma 3.3. *For all $x \in \mathbb{Z}^d$, $d \geq 3$, and $R > r > 0$ such that $x \in B(R) \setminus B(r)$, we have*

$$\mathbb{P}_x[\tau_{\partial B(R)}^+ < \tau_{B(r)}^+] = \frac{r^{-(d-2)} - \|x\|^{-(d-2)} + O(r^{-(d-1)})}{r^{-(d-2)} - R^{-(d-2)} + O(r^{-(d-1)})}, \tag{3.6}$$

as $r \to \infty$.

Proof This comes out of an application of the optional stopping theorem to the martingale $G(S_{n \wedge \tau_0})$. Indeed, let us abbreviate by p the probability in the left-hand side of (3.5); also, let $g_\downarrow$ and $g_\uparrow$ be the (conditional) expected

---

[5] Well, not really (observe that the simple random walk has period two), but you understand what I mean.

[6] Recall that we used a very similar heuristic argument in the beginning of Section 2.1.

values of $G(S)$ on first-hitting the inner and the outer boundaries of the annulus, i.e.,

$$g_\downarrow = \mathbb{E}_x(G(S_{\tau^+_{B(r)}}) \mid \tau^+_{B(r)} < \tau^+_{\partial B(R)}),$$
$$g_\uparrow = \mathbb{E}_x(G(S_{\tau^+_{\partial B(R)}}) \mid \tau^+_{\partial B(R)} < \tau^+_{B(r)}).$$

Since $S_0 = x$, the optional stopping theorem (Corollary 1.7) implies that

$$G(x) = G(S_0) = \mathbb{E}_x G(S_{\tau_{B(r) \cup \partial B(R)}}) = p g_\uparrow + (1-p) g_\downarrow,$$

meaning that

$$p = \frac{g_\downarrow - G(x)}{g_\downarrow - g_\uparrow}. \tag{3.7}$$

Note that $S_{\tau^+_{B(r)}} \in \partial B(r)$ because $x \notin B(r)$; since for any $y \in \partial B(h)$ it holds that $h - 1 < \|y\| \le h$, we have by (3.5) that

$$g_\downarrow = \frac{\gamma_d}{r^{d-2}}(1 + O(r^{-1})),$$

and

$$g_\uparrow = \frac{\gamma_d}{R^{d-2}}(1 + O(R^{-1})).$$

Plugging (3.5) and the preceding into (3.7), we obtain (3.6). $\qquad\square$

Sending R to infinity in (3.6), we obtain that, for any $x \notin B(r)$,

$$\mathbb{P}_x[\tau_{B(r)} = \infty] = 1 - \frac{\|x\|^{-(d-2)} + O(r^{-(d-1)})}{r^{-(d-2)} + O(r^{-(d-1)})}$$
$$= 1 - \left(\frac{r}{\|x\|}\right)^{d-2} + O(r^{-1}). \tag{3.8}$$

Let us now move on to the next fundamental notion of the (discrete) potential theory.

Capacity

For finite $A \subset \mathbb{Z}^d$ and $x \in \mathbb{Z}^d$, let us denote

$$\mathrm{Es}_A(x) = \mathbb{P}_x[\tau^+_A = \infty]\mathbf{1}\{x \in A\}. \tag{3.9}$$

By definition, this quantity is 0 outside A, and, at $x \in A$, it is the *escape probability* from A. Note that $\mathrm{Es}_A(x)$ can be positive only if $x \in \partial A$.

The capacity of a finite set $A \subset \mathbb{Z}^d$ is defined by

$$\mathrm{cap}(A) = \sum_{x \in A} \mathrm{Es}_A(x); \tag{3.10}$$

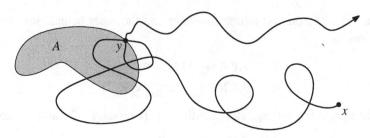

Figure 3.1 On the proof of (3.11).

clearly, it holds that the capacity is translation invariant, and cap(A) = cap(∂A).

What is this notion of capacity good for? To answer this question, we need some preparatory steps. Consider a finite $A \subset \mathbb{Z}^d$. Let us prove a relation that will be used many times later in this chapter: for any $y \in \mathbb{Z}^d$, it holds that

$$\mathbb{P}_x[\tau_A < \infty] = \sum_{y \in A} G(x, y) \operatorname{Es}_A(y) = \sum_{y \in \mathbb{Z}^d} G(x, y) \operatorname{Es}_A(y). \tag{3.11}$$

For the proof, we use an important idea called *the last-visit decomposition*. On the event $\{\tau_A < \infty\}$, let

$$\sigma = \max\{n : S_n \in A\}$$

be the moment of the *last* visit to A (if the walk did not hit A at all, just set σ to be 0). By transience (recall that A is finite!), it is clear that σ is a.s. finite. It is important to note that σ is *not* a stopping time, which actually turns out to be good! To understand why, let us first observe that, by the strong Markov property, the walk's trajectory after any stopping time is "free", that is, it just behaves as a simple random walk starting from the position it had at that stopping time. Now, if we know that σ happened at a given moment, then we know something about the future, namely, we know that the walk must not return to A anymore. In other words, after σ the walk's law is the *conditioned* (on $\tau_A^+ = \infty$) one.[7] Now, look at Figure 3.1: what is the probability that the walker visits $y \in A$ exactly k times (on the picture, $k = 2$), and then escapes to infinity, being y the last visited point of A?

This probability is the total weight of the trajectories such that first they visit y exactly k times and then escape to infinity not touching A anymore.

---

[7] Strictly speaking, this statement needs a rigorous proof; we leave it to the reader as an exercise.

This means that for any $y \in A$ and $k \geq 1$, it holds that

$$\mathbb{P}_x[\text{exactly } k \text{ visits to } y, S_\sigma = y] = \mathbb{P}_x[\text{at least } k \text{ visits to } y] \operatorname{Es}_A(y). \quad (3.12)$$

Okay, maybe, at first sight it is not clear why "exactly k visits" in the left-hand side became "at least k visits" in the right-hand side. To understand this, think again of the piece of the trajectory till σ. If we consider *only* such trajectories, they correspond to the event $\{y$ is visited *at least* k times$\}$ – indeed, if we only observe the trajectory till the kth visit, we then know that this last event occurred.

Then, summing (3.12) in k from 1 to ∞, we obtain[8]

$$\mathbb{P}_x[\tau_A < \infty, S_\sigma = y] = G(x, y) \operatorname{Es}_A(y), \quad (3.13)$$

and summing (3.13) in $y \in A$, we obtain (3.11).

Now we are able to obtain the following useful corollary of (3.11):

$$\operatorname{cap}(A) \min_{z \in A} G(x, z) \leq \mathbb{P}_x[\tau_A < \infty] \leq \operatorname{cap}(A) \max_{z \in A} G(x, z); \quad (3.14)$$

informally, at least in the case when $\min_{z \in A} \|x - z\|$ and $\max_{z \in A} \|x - z\|$ are of the same order, (3.14) means that the probability of ever hitting the set A is proportional to its capacity, and (by (3.5)) is inversely proportional to the distance to that set to power $d - 2$. This justifies the (already mentioned) intuition that the capacity measures how large the set is from the point of view of the simple random walk.

Next, let us obtain the exact expressions (in terms of the Green function G) for capacities of one- and two-point sets. We are going to prove that

$$\operatorname{cap}(\{0\}) = \frac{1}{G(0)}, \quad (3.15)$$

$$\operatorname{cap}(\{0, x\}) = \frac{2}{G(0) + G(x)} \quad (3.16)$$

(by translation invariance, the preceding equations also yield the expressions for $\operatorname{cap}(\{x\})$ and $\operatorname{cap}(\{x, y\})$). Indeed, first, under $\mathbb{P}_0$, the number of visits to the origin (counting the "initial" one at time 0) is a geometric random variable with success probability $\operatorname{Es}_{\{0\}}(0)$. So, its mean is $1/\operatorname{Es}_{\{0\}}(0)$ on one hand and $G(0)$ on the other hand, meaning that $\operatorname{Es}_{\{0\}}(0) = 1/G(0)$. Since, by definition, $\operatorname{cap}(\{0\}) = \operatorname{Es}_{\{0\}}(0)$, we obtain (3.15). The argument for two-point sets is very similar: let $p := \operatorname{Es}_{\{0,x\}}(0)$; by symmetry, it holds also that $p = \operatorname{Es}_{\{0,x\}}(x)$. So, the total number of visits to $\{0, x\}$ (starting from

---

[8] Recall that, for a nonnegative integer-valued random variable ζ, it holds that
$\mathbb{E}\zeta = \sum_{k \geq 1} \mathbb{P}[\zeta \geq k]$.

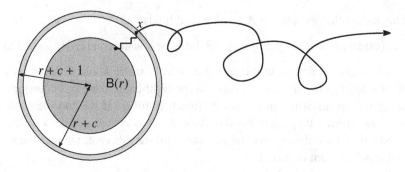

Figure 3.2 Escaping from a ball.

either 0 or x) has geometric distribution with success probability p, meaning that $p^{-1} = G(0) + G(x)$. Since $\mathrm{cap}(\{0, x\}) = \mathrm{Es}_{\{0,x\}}(0) + \mathrm{Es}_{\{0,x\}}(x) = 2p$, (3.16) follows.

The preceding argument can be also used to calculate the capacities of other *symmetric* sets, like e.g. the four vertices of a "spacial rectangle" or other similar things.

As for the capacity of a d-dimensional (discrete) ball, let us show that (with the constant γ_d from (3.5))

$$\mathrm{cap}(\mathsf{B}(r)) \sim \frac{r^{d-2}}{\gamma_d} \qquad \text{as } r \to \infty. \tag{3.17}$$

To *understand* why $\mathrm{cap}(\mathsf{B}(r))$ should be of order r^{d-2}, consider x such that $\|x\| \in [r+c, r+c+1]$, where $c > 0$ is a large enough constant; see Figure 3.2. Note that (3.8) yields

$$\begin{aligned}
\mathbb{P}_x[\tau^+_{\mathsf{B}(r)} = \infty] &= 1 - \left(\frac{r}{\|x\|}\right)^{d-2} + O(r^{-1}) \\
&= 1 - \left(1 + \frac{\|x\| - r}{r}\right)^{-(d-2)} + O(r^{-1}) \\
&= -\frac{(d-2)c}{r} + O(r^{-1})
\end{aligned}$$

(where there is no dependence on c in the O's). The two terms in the preceding expression are of the same order, but we are allowed to make c as large as we want; this will imply that, for such x, $\mathbb{P}_x[\tau^+_{\mathsf{B}(r)} = \infty] \asymp r^{-1}$. This by its turn means that $\mathrm{Es}_{\mathsf{B}(r)}(y) \asymp r^{-1}$ (observe that, clearly, from any boundary point of $\mathsf{B}(r)$ it is possible to walk to some x as before with uniformly positive probability). Since $|\partial \mathsf{B}(r)| \asymp r^{d-1}$, we see that the capacity of $\mathsf{B}(r)$ is indeed of order $r^{d-1} \times r^{-1} = r^{d-2}$.

To obtain the more precise relation (3.17), we need the following.

Proposition 3.4. *For any finite* $A \subset \mathbb{Z}^d$, *it holds that*

$$\text{cap}(A) = \lim_{x \to \infty} \frac{\mathbb{P}_x[\tau_A < \infty]}{G(x)} = \lim_{x \to \infty} \frac{\|x\|^{d-2}}{\gamma_d} \mathbb{P}_x[\tau_A < \infty]. \qquad (3.18)$$

Proof This is a direct consequence of (3.14) and (3.5): note that as $x \to \infty$, both $\min_{z \in A} G(x, z)$ and $\max_{z \in A} G(x, z)$ are asymptotically equivalent to $\gamma^d \|x\|^{-(d-2)}$. $\qquad \square$

It remains to observe that the asymptotics of the capacity of a ball (relation (3.17)) follows from (3.8) and Proposition 3.4.

We are now going to present an interesting application of the technique we just developed. Let us recall the following

Definition 3.5. We say that $A \subset \mathbb{Z}^d$ is *recurrent*, if $\mathbb{P}_x[\tau_A < \infty] = 1$ for all $x \in \mathbb{Z}^d$. Otherwise, we call the set A *transient*.

Clearly, the question if a *set* is recurrent or transient is, in principle, not so trivial. As a start, we obtain the following result:

Proposition 3.6. *If* A *is recurrent, then*

$$\sum_{k=1}^{\infty} 1\{S_k \in A\} = \infty \qquad \mathbb{P}_x\text{-a.s.},$$

for all $x \in \mathbb{Z}^d$, *that is, regardless of the starting point,* A *is visited infinitely many times a.s.. If* A *is transient, then*

$$\sum_{k=1}^{\infty} 1\{S_k \in A\} < \infty \qquad \mathbb{P}_x\text{-a.s.},$$

for all $x \in \mathbb{Z}^d$.

Proof The first part (that recurrence implies that A is visited infinitely many times a.s.) is evident; let us prove the second part. Let A be a transient set and let us define the function h by

$$h(x) = \mathbb{P}_x[A \text{ is visited infinitely often}].$$

An immediate observation is that h is harmonic – just use the formula of total probability, conditioning on the first step. So, since h is also obviously bounded, Proposition 3.1 implies that this function is constant, $h(x) = p \in [0, 1]$ for all x.

Now, what can be the value of p? First, it has to be strictly less than 1 by transience of A (there is at least one site x_0 such that $\mathbb{P}_{x_0}[\tau_A < \infty] < 1$ and, obviously, $h(x_0) \leq \mathbb{P}_{x_0}[\tau_A < \infty]$). Next, write (conditioning on the first entrance to A, if any)

$$
\begin{aligned}
p &= h(x_0) \\
&= \sum_{y \in A} \mathbb{P}_{x_0}[\tau_A < \infty, S_{\tau_A} = y]h(y) \\
&= p \sum_{y \in A} \mathbb{P}_{x_0}[\tau_A < \infty, S_{\tau_A} = y] \\
&= p\,\mathbb{P}_{x_0}[\tau_A < \infty],
\end{aligned}
$$

and (since, as we just assumed, $\mathbb{P}_{x_0}[\tau_A < \infty] < 1$) this implies that $p = 0$.   □

As we know already, in dimensions $d \geq 3$ the one-point sets are transient; Proposition 3.6 then implies[9] that all finite sets are transient as well. But what can we say about infinite sets? The answer is given by the following theorem:

Theorem 3.7 (Wiener's criterion). *For $d \geq 3$, $A \subset \mathbb{Z}^d$ is recurrent if and only if*

$$
\sum_{k=1}^{\infty} \frac{\operatorname{cap}(A_k)}{2^{(d-2)k}} = \infty, \tag{3.19}
$$

where

$$
A_k = \{x \in A : 2^{k-1} < \|x\| \leq 2^k\}
$$

is the intersection of A with the annulus $\mathsf{B}(2^k) \setminus \mathsf{B}(2^{k-1})$.

The proof is left to the reader (Exercises 3.13 and 3.14).

The last of the three main ingredients that make our naive view of the potential theory is harmonic measure; we start looking at it now.

Harmonic measure

As before, let A be a finite subset of $\mathbb{Z}^d$, $d \geq 3$. The harmonic measure $\operatorname{hm}_A(\cdot)$ on A is defined by

$$
\operatorname{hm}_A(x) = \frac{\operatorname{Es}_A(x)}{\operatorname{cap}(A)}, \quad x \in A, \tag{3.20}
$$

---

[9] How exactly?

that is, the value of $\text{hm}_A(x)$ is proportional to the escape probability from x to infinity. Remarkably, it is also true that hm_A is the "entrance measure to A from infinity", that is, the following result holds:

Theorem 3.8. *For all $y \in A$, we have*

$$\text{hm}_A(y) = \lim_{x \to \infty} \mathbb{P}_x[S_{\tau_A} = y \mid \tau_A < \infty]. \tag{3.21}$$

Why should Theorem 3.8 be valid? Let us first give an informal explanation. Consider $y, z \in \partial A$, such that $y \neq z$ and both $\text{Es}_A(y)$ and $\text{Es}_A(z)$ are strictly positive. Then the ratio of the "total weights" of trajectories which escape A from, respectively, y and z, equals $\text{Es}_A(y)/\text{Es}_A(z)$. Now, if x is very far away from A and the walker that started somewhere at A happens to pass through x, it likely does not "remember" its exact starting position. Since the time reversal does not change the "weight" of the trajectory,[10] the ratio of the chances that a trajectory passing through x will end up in y (respectively, in z) should be then $\text{Es}_A(y)/\text{Es}_A(z)$ as well.

Now, the rigorous proof of the preceding result may look not very intuitive at first sight, but note that it also makes use of this reversibility property.

Proof of Theorem 3.8. We now use a trajectory-counting argument very similar to the one in the proof of (3.11). For $x \notin A$, $y \in \partial A$, and $n \geq 1$, let us denote by $\Theta_{xy}^{(n)}$ the set of nearest-neighbour trajectories $\wp = (z_0, \ldots, z_k)$ such that

- $z_0 = x$, $z_k = y$, and $z_j \notin A$ for all $j \leq k - 1$, i.e., the trajectory ends on the first entrance to A, which takes place in y;
- $\sum_{j=0}^{k} \mathbf{1}\{z_j = x\} = n$, i.e., the trajectory visits x exactly n times (note that we *do* count $z_0 = x$ as one visit);

see Figure 3.3. For such trajectory, we also write $|\wp| = k$ to denote its length, and $P_\wp = (2d)^{-|\wp|}$ to denote its weight (a.k.a. probability). Let us also denote by

$$N_x = \sum_{j=0}^{\infty} \mathbf{1}\{S_j = x\}$$

the total number of visits to $x \notin A$, by

$$N_x^b = \sum_{j=0}^{\tau_A^+ - 1} \mathbf{1}\{S_j = x\}$$

---

[10] Formally, for infinite trajectories this only means that $0 = 0$, but you understand what I wanted to say.

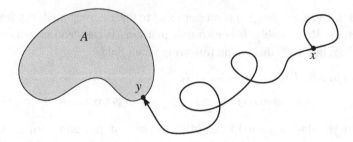

Figure 3.3 On the proof of Theorem 3.8: an example of a trajectory from $\Theta_{xy}^{(2)}$.

the number of visits to x *before* the first return to A, and by

$$N_x^\sharp = \sum_{j=\tau_A^+}^\infty \mathbf{1}\{S_j = x\}$$

the number of visits to x *after* the first return to A (naturally, setting $N_x^\sharp = 0$ on $\{\tau_A^+ = \infty\}$).

It is clear that

$$\mathbb{P}_x[\tau_A < \infty, S_{\tau_A} = y] = \sum_{n=1}^\infty \sum_{\wp \in \Theta_{xy}^{(n)}} P_\wp \tag{3.22}$$

(we just sum the weights of all trajectories starting at x and entering A at y). The next relation may seem a bit less clear, but it is here where we use the reversibility property:

$$\mathbb{P}_y[N_x^\flat \geq n] = \sum_{\wp \in \Theta_{xy}^{(n)}} P_\wp. \tag{3.23}$$

Indeed (quite analogously to the proof of (3.11)), when, starting at y, we see a reversal of a trajectory from $\Theta_{xy}^{(n)}$, we are sure that the event $\{N_x^\flat \geq n\}$ occurs. Therefore, we can write

$$\mathbb{P}_x[S_{\tau_A} = y \mid \tau_A < \infty]$$

$$= \frac{\mathbb{P}_x[\tau_A < \infty, S_{\tau_A} = y]}{\mathbb{P}_x[\tau_A < \infty]}$$

(by (3.22) and (3.23))

$$= (\mathbb{P}_x[\tau_A < \infty])^{-1} \sum_{n=1}^\infty \mathbb{P}_y[N_x^\flat \geq n]$$

$$= (\mathbb{P}_x[\tau_A < \infty])^{-1} \mathbb{E}_y N_x^{\flat}$$

(since, clearly, $N_x = N_x^{\flat} + N_x^{\sharp}$)

$$= (\mathbb{P}_x[\tau_A < \infty])^{-1} (\mathbb{E}_y N_x - \mathbb{E}_y N_x^{\sharp})$$

(conditioning on the position of the first re-entry to A)

$$= (\mathbb{P}_x[\tau_A < \infty])^{-1} \Big(G(y, x) - \sum_{z \in \partial A} \mathbb{P}_y[\tau_A^+ < \infty, S_{\tau_A^+} = z] G(z, x) \Big). \qquad (3.24)$$

Then Proposition 3.4 together with (3.5) imply that, for any fixed $z \in \mathbb{Z}^d$,

$$\frac{G(z, x)}{\mathbb{P}_x[\tau_A < \infty]} \to \frac{1}{\operatorname{cap}(A)} \quad \text{as } x \to \infty.$$

So, sending x to infinity in (3.24), we obtain

$$\lim_{x \to \infty} \mathbb{P}_x[S_{\tau_A} = y \mid \tau_A < \infty] = \frac{1}{\operatorname{cap}(A)} \Big(1 - \sum_{z \in \partial A} \mathbb{P}_y[\tau_A^+ < \infty, S_{\tau_A^+} = z] \Big)$$

$$= \frac{1}{\operatorname{cap}(A)} (1 - \mathbb{P}_y[\tau_A^+ < \infty])$$

$$= \frac{\mathbb{P}_y[\tau_A^+ = \infty]}{\operatorname{cap}(A)}$$

$$= \operatorname{hm}_A(y),$$

thus concluding the proof of Theorem 3.8. $\qquad \square$

Many other interesting things can be said about the transient case, but we prefer to stop here and pass to the recurrent one. [11]

3.2 Potential theory in two dimensions

In this section, we try to do roughly the same as in the previous one, only in two dimensions. As we know, there is one big difference between the dimension two and higher dimensions: as shown in Chapter 2, unlike the higher-dimensional SRW, the two-dimensional SRW is recurrent. This means that the mean number of visits from any site to any other site equals infinity; this prevents us from defining Green's function in the same way as in Chapter 3.1. In spite of this unfortunate circumstance, we still would like to use martingale arguments, so a "substitute" of Green's function is needed. Now, here comes the key observation: while the mean number of visits to the origin is infinite, the *difference* between the mean number of

---

[11] Revenons à nos moutons ©.

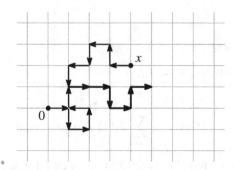

Figure 3.4 The coupling of two random walks starting at 0 and $x = (4, 2)$. Note that their first coordinates become equal at time 3 (when the first walk is at $(2, -1)$ and the second one is at $(2, 3)$), and the walks meet at time 6 at site $(1, 1)$.

visits to the origin starting from 0 and starting from x is finite, *if suitably defined*. Let us do it now.

Potential kernel

Namely, let us define the *potential kernel* $a(\cdot)$ by

$$a(x) = \sum_{k=0}^{\infty} (\mathbb{P}_0[S_k = 0] - \mathbb{P}_x[S_k = 0]), \qquad x \in \mathbb{Z}^2. \qquad (3.25)$$

By definition, it holds that $a(0) = 0$. To see that the limit (finite or infinite) in (3.25) actually exists is a little bit more subtle, but still quite elementary. Indeed, Exercise 3.28 (iii) implies that

- if x is an even site (i.e., the sum of its coordinates is even), then all terms in the summation (3.25) are nonnegative (more precisely, they are positive for even k and zero for odd k);
- if x is an odd site, then we have a series with alternating signs in (3.25), but the sums of each two consecutive terms (i.e., $\mathbb{P}_0[S_{2k} = 0] - \mathbb{P}_x[S_{2k+1} = 0]$) are again strictly positive and converge to zero, which clearly implies that the sum is well defined.

Note that the preceding argument also implies that $a(x) > 0$ for all $x \neq 0$.

Now, let us convince ourselves that the series converges (i.e., $a(x)$ is finite) for all $x \in \mathbb{Z}^2$, and figure out how large $a(x)$ should be. The "normal" approach would be employing the *local* Central Limit Theorem[12] for this,

---

[12] Analogous to the De Moivre-Laplace one, only in two dimensions; see e.g. theorem 2.1.1 of [63].

but we prefer to use another interesting and very useful tool called *coupling*.[13] It will be a *long* argument (sorry for that!) since we will need to do a three-parts divide-and-conquer argument (later it will be clear what is meant by this), but it is still nice and instructive. Assume that both coordinates of $x \neq 0$ are even, so, in particular, two random walks simultaneously started at x and at the origin *can* meet. Next, we construct these two random walks *together*, that is, on the same probability space. We do this in the following way: we first choose one of the two coordinates at random, and then make the walks jump in the opposite directions if the values of the chosen coordinates of the two walks are different, and in the same direction in case they are equal; see Figure 3.4. Formally, assume that, at a given moment n the positions of the walks are S'_n and S''_n; we have then $S'_0 = 0$, $S''_0 = x$. Let $\mathcal{J}_n$ and Z_n be independent random variables assuming values in $\{1, 2\}$ and $\{-1, 1\}$ respectively, with equal (to $\frac{1}{2}$) probabilities. Then, we set:

$$(S'_{n+1}, S''_{n+1}) = \begin{cases} (S'_n + Z_n e_{\mathcal{J}_n}, S''_n - Z_n e_{\mathcal{J}_n}), & \text{if } S'_n \cdot e_{\mathcal{J}_n} \neq S''_n \cdot e_{\mathcal{J}_n}, \\ (S'_n + Z_n e_{\mathcal{J}_n}, S''_n + Z_n e_{\mathcal{J}_n}), & \text{if } S'_n \cdot e_{\mathcal{J}_n} = S''_n \cdot e_{\mathcal{J}_n}. \end{cases}$$

Note that if the first (second) coordinates of the two walks are equal at some moment, then they will remain so forever. This means, in particular, that, when the two walks meet, they stay together. Let us assume, for definiteness, that x belongs to the first quadrant, that is, $x = (2b_1, 2b_2)$ for $b_{1,2} \geq 0$. Let

$$T_j = \min\{n \geq 0 : S'_n \cdot e_j = S'_n \cdot e_j\}$$

for $j = 1, 2$; that is, T_j is the moment when the jth coordinates of S' and S'' coincide for the first time. Notice that, alternatively, one can express them in the following way:

$$T_j = \min\{n \geq 0 : S'_n \cdot e_j = b_j\} = \min\{n \geq 0 : S''_n \cdot e_j = b_j\} \qquad (3.26)$$

(clearly, they have to meet exactly in the middle). Let also $T = T_1 \vee T_2$ be the *coupling time*, i.e., the moment when the two walks meet and stay together.

Now we go back to (3.25) and use the strategy usually called "divide and conquer": write

$$a(x) = \sum_{k < \|x\|} (\mathbb{P}_0[S_k = 0] - \mathbb{P}_x[S_k = 0])$$

---

[13] We used it already (without calling it so) in the proof of Proposition 3.1; here, we take it further.

$$+ \sum_{k \in [\|x\|, \|x\|^3]} (\mathbb{P}_0[S_k = 0] - \mathbb{P}_x[S_k = 0])$$

$$+ \sum_{k > \|x\|^3} (\mathbb{P}_0[S_k = 0] - \mathbb{P}_x[S_k = 0])$$

$$=: M_1 + M_2 + M_3,$$

and then let us deal with the three terms separately.

First, let us recall the calculations from Section 2.1: we have obtained there that

$$\mathbb{P}_0[S_{2k} = 0] \asymp \frac{1}{k}. \tag{3.27}$$

To deal with the term M_1, just observe that $\mathbb{P}_x[S_k = 0] = 0$ for $k < \|x\|$ — there is simply not enough time for the walker to go from x to 0. The relation (3.27) then implies that

$$M_1 \asymp \ln \|x\|. \tag{3.28}$$

For the second term, we have already observed that all summands there are nonnegative, so (3.27) implies that

$$0 \le M_2 \lesssim \ln \|x\|. \tag{3.29}$$

That is, M_1 is of order $\ln \|x\|$, and M_2 is nonnegative and *at most* of order $\ln \|x\|$; this clearly implies that the sum of them is also of order $\ln \|x\|$.

It remains to deal with the term M_3. It is here that we use the coupling idea: let us write

$$\sum_{k > \|x\|^3} (\mathbb{P}_0[S_k = 0] - \mathbb{P}_x[S_k = 0])$$

$$= \mathbb{E} \sum_{k > \|x\|^3} (\mathbf{1}\{S'_k = 0\} - \mathbf{1}\{S''_k = 0\})$$

(write $1 = \mathbf{1}\{T \le k\} + \mathbf{1}\{T > k\}$, and note that the kth term is 0 on $\{T \le k\}$)

$$= \mathbb{E} \sum_{k > \|x\|^3} (\mathbf{1}\{S'_k = 0\} - \mathbf{1}\{S''_k = 0\})\mathbf{1}\{T > k\}$$

(if $T > k$, then S''_k cannot be at the origin; recall (3.26))

$$= \mathbb{E} \sum_{k > \|x\|^3} \mathbf{1}\{S'_k = 0\}\mathbf{1}\{T > k\}$$

$$= \sum_{k > \|x\|^3} \mathbb{P}[S'_k = 0, T > k]$$

(since $\{T > k\} = \{T_1 > k\} \cup \{T_2 > k\}$)

$$\leq \sum_{k > \|x\|^3} \mathbb{P}[S'_k = 0, T_1 > k] + \sum_{k > \|x\|^3} \mathbb{P}[S'_k = 0, T_2 > k]$$

(by symmetry)

$$= 2 \sum_{k > \|x\|^3} \mathbb{P}[S'_k = 0, T_1 > k]. \tag{3.30}$$

We are now going to prove that the terms in the preceding sum are at most of order $k^{-4/3}$. For this, we first prove that, for all $m \geq b_1^3$,

$$\mathbb{P}_0[X_{2m} = 0, \widehat{T}(b_1) > 2m] \leq m^{-5/6}, \tag{3.31}$$

where X is a one-dimensional simple random walk, and $\widehat{T}(s) = \min\{\ell > 0 : X_\ell = s\}$. To show (3.31), we use the following well known fact:

Proposition 3.9 (The Reflection Principle). *Let us consider oriented paths*[14] *in $\mathbb{Z}^2$, such that from x the path can go to either to $x + e_1 - e_2$ or to $x + e_1 + e_2$ (that is, it can go to northeast and southeast directions). Let two sites $x = (x_1, x_2)$ and $y = (y_1, y_2)$ be such that $x_1 < y_1, x_2 > 0, y_2 > 0$. Then the number of paths that go from x to y and have at least one common point with the horizontal axis is equal to the* total *number of paths that go from $\tilde{x} = (x_1, -x_2)$ to y.*

Proof   Just look at Figure 3.5 (on the right). In a way, it is the same coupling as before, only in dimension 1.   □

Now, we write

$$\mathbb{P}_0[X_{2m} = 0, \widehat{T}(b_1) \leq 2m] = \mathbb{P}_{b_1}[X_{2m} = b_1, \widehat{T}(0) \leq 2m]$$

(by the Reflection Principle)

$$= \mathbb{P}_{-b_1}[X_{2m} = b_1]$$

$$= 2^{-2m}\binom{2m}{m - b_1}.$$

So, we have for $m \geq b_1^3$,

$$\mathbb{P}_0[X_{2m} = 0, \widehat{T}(b_1) > 2m]$$

$$= 2^{-2m}\left(\binom{2m}{m} - \binom{2m}{m - b_1}\right)$$

---

[14]   These are the *space-time* paths of one-dimensional simple random walk, the horizontal axis represents time and the vertical axis represents space.

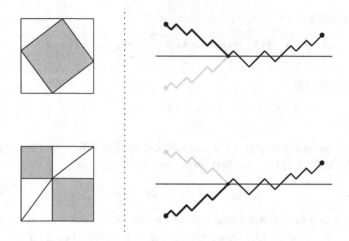

Figure 3.5 Two famous visual proofs.

$$= 2^{-2m}\binom{2m}{m}\Big(1 - \frac{(m - b_1 + 1)\cdots(m - 1)m}{(m + 1)\cdots(m + b_1)}\Big)$$

$$= 2^{-2m}\binom{2m}{m}\Big(1 - \Big(1 - \frac{b_1}{m + 1}\Big)\cdots\Big(1 - \frac{b_1}{m + b_1}\Big)\Big)$$

$$\leq 2^{-2m}\binom{2m}{m}\Big(1 - \Big(1 - \frac{b_1}{m}\Big)^{b_1}\Big)$$

(using the simple inequality $(1 - x)^a \geq 1 - ax$ for $x \in [0, 1]$ and $a \geq 1$)

$$\leq 2^{-2m}\binom{2m}{m} \times \frac{b_1^2}{m}$$

(recall the calculation (2.2) from Section 2.1 and use that $b_1^2 \leq m^{2/3}$)

$$\leq \frac{1}{m^{1/2}} \times \frac{m^{2/3}}{m}$$

$$= \frac{1}{m^{5/6}},$$

thus proving (3.31). Note also that, if B_k is a Binomial$(2k, \frac{1}{2})$ random variable (think of the number of steps in vertical direction of two-dimensional simple random walk up to time $2k$), then it holds that[15]

$$\mathbb{P}[\tfrac{2}{3}k \leq B_k \leq \tfrac{4}{3}k] = \mathbb{P}[|B_k - \mathbb{E}B_k| \leq \tfrac{1}{3}k] \geq 1 - e^{-ck}. \qquad (3.32)$$

[15] Use e.g. Chernoff's bound.

So, we have

$$\mathbb{P}[S'_{2k} = 0, T_1 > 2k]$$

$$= \sum_{m=0}^{k} \mathbb{P}[B_k = m]\mathbb{P}_0[X_{2m} = 0, \widehat{T}(b_1) > 2m]\mathbb{P}_0[X_{2(k-m)} = 0]$$

(due to (3.32), only the "middle terms" matter, and they are all of the same order)

$$\lesssim \frac{1}{k^{5/6}} \times \frac{1}{k^{1/2}}$$

$$= k^{-4/3};$$

going back to (3.30), we find that the term M_3 is bounded above by a constant (in fact, even polynomially small in $\|x\|$), and this finally shows that, for $x \in \mathbb{Z}^2$ with both coordinates even, $a(x)$ exists and is of order $\ln\|x\|$. We are finally done with the long proof of existence, but the reader is advised to see Exercise 3.29.

The preceding argument can be tweaked to treat *all* $x \in \mathbb{Z}^2$, but, as we will see now, this is unnecessary. Let us show that the function a is harmonic outside the origin, i.e.,

$$a(x) = \frac{1}{4}\sum_{y \sim x} a(y) \quad \text{for all } x \neq 0. \tag{3.33}$$

Then, assuming that (3.33) holds, it is clear that the fact that $a(x) < \infty$ for *some* $x \neq 0$ implies that $a(x) < \infty$ for *all* $x \in \mathbb{Z}^2$ (indeed, if $a(x) < \infty$ then $a(y)$ should also be finite for all $y \sim x$, etc.). Also, if $x, y \neq 0$ and $y \sim x$, then (3.33) implies that $\frac{1}{4}a(x) \leq a(y) \leq 4a(x)$, and this shows that $a(x)$ should be of order $\ln\|x\|$ as $x \to \infty$ indeed.

Now, we prove (3.33). This is again a consequence of the total expectation formula; we only need to take some more care this time because of the limits involved. Let

$$N_z^{(k)} = \sum_{j=0}^{k} \mathbf{1}\{S_k = z\} \tag{3.34}$$

be the number of visits to z up to time k; with this notation, we have $a(x) = \lim_{k\to\infty}(\mathbb{E}_0 N_0^{(k)} - \mathbb{E}_x N_0^{(k)})$. The total expectation formula (conditional on the first step) gives us that, for $x \neq 0$,

$$\mathbb{E}_x N_0^{(k)} = \frac{1}{4}\sum_{y \sim x} \mathbb{E}_y N_0^{(k-1)},$$

so

$$\mathbb{E}_0 N_0^{(k)} - \mathbb{E}_x N_0^{(k)} = \mathbb{E}_0 N_0^{(k)} - \frac{1}{4} \sum_{y \sim x} \mathbb{E}_y N_0^{(k-1)}$$

$$= \mathbb{E}_0 N_0^{(k-1)} + \mathbb{P}_0[S_k = 0] - \frac{1}{4} \sum_{y \sim x} \mathbb{E}_y N_0^{(k-1)}$$

$$= \mathbb{P}_0[S_k = 0] + \frac{1}{4} \sum_{y \sim x} (\mathbb{E}_0 N_0^{(k-1)} - \mathbb{E}_y N_0^{(k-1)}).$$

Sending k to infinity in the preceding, we obtain (3.33).

Let $\mathcal{N} = \{\pm e_i, i = 1, 2\}$ be the set of the four neighbours of the origin. Another useful fact is that

$$a(x) = 1 \text{ for all } x \in \mathcal{N}. \tag{3.35}$$

To see this, first, observe that, by symmetry, $a(\cdot)$ must have the same value on the sites of $\mathcal{N}$. Then, again by the total expectation formula,

$$\mathbb{E}_0 N_0^{(k)} = 1 + \frac{1}{4} \sum_{y \in \mathcal{N}} \mathbb{E}_y N_0^{(k-1)} = 1 + \mathbb{E}_{e_1} N_0^{(k-1)}$$

(note that the time-zero visit counts, and then use symmetry). The preceding implies that

$$\mathbb{E}_0 N_0^{(k)} - \mathbb{E}_{e_1} N_0^{(k)} = (1 + \mathbb{E}_{e_1} N_0^{(k-1)}) - (\mathbb{E}_{e_1} N_0^{(k-1)} + \mathbb{P}_{e_1}[S_k = 0])$$

$$= 1 - \mathbb{P}_{e_1}[S_k = 0],$$

and, sending k to infinity, we obtain (3.35).

Again, as with Green's function in the previous section (expression (3.5)), we argue that the common wisdom *suggests* that the potential kernel should be roughly spherically symmetric, and "well behaved" in general. Indeed, it is possible to prove that, as $x \to \infty$,

$$a(x) = \frac{2}{\pi} \ln \|x\| + \gamma' + O(\|x\|^{-2}), \tag{3.36}$$

where, being $\gamma = 0.5772156\ldots$ the Euler–Mascheroni constant[16],

$$\gamma' = \frac{2\gamma + \ln 8}{\pi} = 1.0293737\ldots, \tag{3.37}$$

cf. theorem 4.4.4 of [63], and see Exercises 3.31 and 3.32. We also note that it is possible to obtain exact values of $a(\cdot)$ in other sites (close to the

[16] $\gamma = \lim_{n \to \infty} (1 + \frac{1}{2} + \cdots + \frac{1}{n} - \ln n)$.

origin); for example, it holds that $a(e_1 + e_2) = \frac{4}{\pi}$, $a(2e_1) = 4 - \frac{8}{\pi}$, and so on; see section III.15 of [91].

Observe that the harmonicity of a outside the origin (established in (3.33)) immediately implies that the following result holds:

Proposition 3.10. *The process $a(S_{k \wedge \tau_0})$ is a martingale.*

We will repeatedly use this fact in the sequel. What we will also repeatedly use, is that, due to (3.36),

$$a(x + y) - a(x) = O(\tfrac{\|y\|}{\|x\|}) \tag{3.38}$$

for all $x, y \in \mathbb{Z}^2$ such that (say) $\|x\| > 2\|y\|$.

With some (slight) abuse of notation, we also consider the function

$$a(r) = \frac{2}{\pi} \ln r + \gamma'$$

of a *real* argument $r \geq 1$. Note that, in general, $a(x)$ need not be equal to $a(\|x\|)$, although they are of course quite close for large x. The advantage of using this notation is e.g. that, due to (3.36) and (3.38), we may write (for fixed x or at least x such that $2\|x\| \leq r$)

$$\sum_{y \in \partial B(x,r)} v(y)a(y) = a(r) + O(\tfrac{\|x\| \vee 1}{r}) \qquad \text{as } r \to \infty \tag{3.39}$$

for *any* probability measure v on $\partial B(x, r)$.

As in the higher-dimensional case, we need an asymptotic expression for the probability that a random walk in an annulus leaves it through its outer boundary. Quite analogously to the proof of Lemma 3.3, we can obtain the following result from (3.36), Proposition 3.10, and the optional stopping theorem:

Lemma 3.11. *For all $x \in \mathbb{Z}^2$ and $R > r > 0$ such that $x \in B(y, R) \setminus B(r)$, we have*

$$\mathbb{P}_x[\tau_{\partial B(y,R)} < \tau_{B(r)}] = \frac{a(x) - a(r) + O(r^{-1})}{a(R) - a(r) + O(r^{-1} + \frac{\|y\|+1}{R})}, \tag{3.40}$$

as $r, R \to \infty$. The preceding also holds with $\tau_{B(y,R)^c}$ on the place of $\tau_{\partial B(y,R)}$.

We do not write its proof here (since it is *really* analogous); but, in case the reader prefers to see a similar proof again, we prove the following result for the probability of escaping the origin:

Lemma 3.12. *Assume that $x \in B(y, r)$ and $x \neq 0$. Then*

$$\mathbb{P}_x[\tau_{\partial B(y,r)} < \tau_0^+] = \frac{a(x)}{a(r) + O(\frac{\|y\|+1}{r})}, \tag{3.41}$$

as $r \to \infty$. As before, this lemma also holds with $\tau_{B(y,R)^C}$ on the place of $\tau_{\partial B(y,r)}$.

Proof   Indeed, use Proposition 3.10, and the optional stopping theorem to write (recall that $a(0) = 0$)

$$a(x) = \mathbb{P}_x[\tau_{\partial B(y,r)} < \tau_0^+]\mathbb{E}_x(a(S_{\tau_{\partial B(y,r)}}) \mid \tau_{\partial B(y,r)} < \tau_0^+),$$

and then use (3.39).   □

Note that Lemma 3.12 implies that (since, from the origin, on the next step the walk will go to a site in $\mathcal{N}$ where the potential kernel equals 1)

$$\mathbb{P}_0[\tau_{\partial B(r)} < \tau_0^+] = \frac{1}{a(r) + O(r^{-1})} = \left(\frac{2}{\pi} \ln r + \gamma' + O(r^{-1})\right)^{-1}. \tag{3.42}$$

The reader may have recalled that this was the formula that was used in the introduction to calculate the probability of going to the edge of our galaxy before returning to the initial point.

Green's function

Wait, but didn't we agree in the beginning of this section that there is no Green's function in two dimensions? Well, this applies to Green's function in the whole space, but we also can define a "restricted Green's function", which can be still quite useful. Let Λ be a (typically, finite) subset of $\mathbb{Z}^2$. For $x, y \in \Lambda$, let us define

$$G_\Lambda(x, y) = \mathbb{E}_x \sum_{k=0}^{\tau_{\Lambda^C}-1} 1\{S_k = y\} \tag{3.43}$$

to be the mean number of visits to y starting from x before stepping out of Λ. Notice that this definition formally makes sense[17] also in the case when at least one of the arguments is outside of Λ, in which case $G_\Lambda(x, y) = 0$.

This notion is, of course, less convenient than that of Green's function in the whole space, since we (clearly) lose the translation invariance, and

---

[17] With the usual convention that $\sum_{k=0}^{-1} = 0$.

also (apparently) lose the symmetry. It is quite remarkable, however, that, in fact, the symmetry is not lost! Indeed, let us prove that

$$G_\Lambda(x, y) = G_\Lambda(y, x) \quad \text{for any } x, y \in \Lambda. \tag{3.44}$$

We use the usual trick of getting rid of random sums (such as the one in (3.43)):

$$\mathbb{E}_x \sum_{k=0}^{\tau_{\Lambda^C}-1} \mathbb{1}\{S_k = y\} = \mathbb{E}_x \sum_{k=0}^{\infty} \mathbb{1}\{S_k = y, \tau_{\Lambda^C} > k\}$$

$$= \sum_{k=0}^{\infty} \mathbb{P}_x[S_k = y, \tau_{\Lambda^C} > k]$$

and so, to prove (3.44), it is enough to show that

$$\mathbb{P}_x[S_k = y, \tau_{\Lambda^C} > k] = \mathbb{P}_y[S_k = x, \tau_{\Lambda^C} > k]$$

for any k. But this is quite evident: indeed, the number of k-step trajectories that lie fully inside Λ, start at x and end at y, is obviously the same as the number of such trajectories that start at y and end at x.

Next, there is a useful relation connecting the restricted Green's function to the potential kernel:

Theorem 3.13. *Assume that Λ is finite. Then it holds that*

$$G_\Lambda(x, y) = \mathbb{E}_x a(S_{\tau_{\Lambda^C}} - y) - a(x - y). \tag{3.45}$$

Proof First, Proposition 3.10 together with (3.35) imply that the process $a(S_n - y)$ is a *submartingale*: indeed, when the walk is not at y, its expected drift equals zero, while, when it is at y, its value is 0 and will become 1 on the next step (so the expected drift equals 1). With a moment of thought, one can see[18] that the process (recall the notation from (3.34))

$$Y_n = a(S_n - y) - N_y^{(n-1)} \tag{3.46}$$

is a *martingale* – the drift of $a(S_n - y)$ (which is present when the walker visits y) is "compensated" by the increase of the value of N_y. Since $G_\Lambda(x, y) = \mathbb{E}_x N_y^{(\tau_{\Lambda^C}-1)}$, it is enough to apply the optional stopping theorem[19] to the martingale Y_n and the stopping time τ_{Λ^C}:

$$a(x - y) = \mathbb{E}_x Y_0 = \mathbb{E}_x Y_{\tau_{\Lambda^C}} = \mathbb{E}_x a(S_{\tau_{\Lambda^C}} - y) - G_\Lambda(x, y),$$

thus proving (3.45). □

[18] Exercise: prove it formally.
[19] Indeed, item (iii) of Corollary 1.7 works fine here, due to the finiteness of Λ and Exercise 3.30.

We use Theorem 3.13 to obtain a convenient expression for the restricted Green's function in the case $\Lambda = B(R)$, with a large R. Let $x, y \in B(R)$. Note that, for any $z \in \partial_e B(R)$, it holds that

$$a(z - y) = a(R) + O(\tfrac{\|y\|+1}{R})$$

(indeed, analogously to (3.38) just observe that $\big| \|z - y\| - R \big| \le \|y\| + 1$ and use that $\ln \|z - y\| = \ln R + \ln (1 + \tfrac{\|z-y\|-R}{R})$). So, Theorem 3.13 implies that, for $x, y \in B(R)$

$$G_{B(R)}(x, y) = a(R) - a(x - y) + O(\tfrac{1+\|x\|\wedge\|y\|}{R}) \qquad (3.47)$$

(as $\|x - y\| \to \infty$)

$$= \frac{2}{\pi} \ln \frac{R}{\|x - y\|} + O(\tfrac{1+\|x\|\wedge\|y\|}{R} + \tfrac{1}{\|x-y\|^2}) \qquad (3.48)$$

(note that, by the symmetry property (3.44), we can assume without loss of generality that $\|y\| \le \|x\|$, so we can choose "the better term" in the preceding O's).

Harmonic measure

Here (I mean, in two dimensions) we take a different route: we first define and discuss the notion of harmonic measure, and only then pass to that of capacity. For a finite $A \subset \mathbb{Z}^2$ and $x \in \mathbb{Z}^2$ let us define

$$q_A(x) = a(x - y_0) - \mathbb{E}_x a(S_{\tau_A} - y_0), \qquad (3.49)$$

where y_0 is *some* site of A (later we will prove that the value of $q_A(x)$ does not depend on the choice of this y_0). Note that $q_A(x) = 0$ for all $x \in A$ (since $\tau_A = 0$ when the walk starts at $x \in A$); also, the preceding definition is invariant under translations, i.e., $q_{A+z}(x + z) = q_A(x)$ for any $z \in \mathbb{Z}^2$. The importance of this quantity is underlined by the following fact:

Proposition 3.14. *For any finite $A \subset \mathbb{Z}^2$ and any $x, y \in \mathbb{Z}^2$, it holds that*

$$q_A(x) = \lim_{R \to \infty} a(R) \mathbb{P}_x[\tau_{B(y,R)^\complement} < \tau_A] = \frac{2}{\pi} \lim_{R \to \infty} \mathbb{P}_x[\tau_{B(y,R)^\complement} < \tau_A] \ln R. \quad (3.50)$$

Proof We use a martingale argument similar to the proofs of Lemmas 3.3, 3.11, and 3.12; we only need to take a bit more care in order to "separate" the term $\mathbb{E}_x a(S_{\tau_A})$ so that it remains "free from conditioning". Assume without restricting generality that $y_0 = 0$ (otherwise we can just "shift" the origin there) and let us apply the optional stopping theorem[20] to the

---

[20] Formally, to justify its use, Corollary 1.7 (ii) is enough.

martingale $a(S_{n \wedge \tau_0})$ with the stopping time $\tau_A \wedge \tau_{B(y,R)^{\complement}}$:

$$a(x) = \mathbb{E}_x a(S_{\tau_A \wedge \tau_{B(y,R)^{\complement}}})$$

$$= \mathbb{E}_x(a(S_{\tau_{B(y,R)^{\complement}}}) \mathbb{1}\{\tau_{B(y,R)^{\complement}} < \tau_A\}) + \mathbb{E}_x(a(S_{\tau_A}) \mathbb{1}\{\tau_A < \tau_{B(y,R)^{\complement}}\})$$

(writing $\mathbb{1}\{\tau_A < \tau_{B(y,R)^{\complement}}\} = 1 - \mathbb{1}\{\tau_{B(y,R)^{\complement}} < \tau_A\}$ in the second term)

$$= \mathbb{P}_x[\tau_{B(y,R)^{\complement}} < \tau_A] \mathbb{E}_x(a(S_{\tau_{B(y,R)^{\complement}}}) \mid \tau_{B(y,R)^{\complement}} < \tau_A)$$

$$+ \mathbb{E}_x a(S_{\tau_A}) - \mathbb{E}_x(a(S_{\tau_A}) \mathbb{1}\{\tau_{B(y,R)^{\complement}} < \tau_A\})$$

$$= \mathbb{P}_x[\tau_{B(y,R)^{\complement}} < \tau_A] \mathbb{E}_x(a(S_{\tau_{B(y,R)^{\complement}}}) \mid \tau_{B(y,R)^{\complement}} < \tau_A)$$

$$+ \mathbb{E}_x a(S_{\tau_A}) - \mathbb{P}_x[\tau_{B(y,R)^{\complement}} < \tau_A] \mathbb{E}_x(a(S_{\tau_A}) \mid \tau_{B(y,R)^{\complement}} < \tau_A),$$

so, abbreviating $b = 1 + \max_{x \in A} \|x\|$, we obtain that

$$\mathbb{P}_x[\tau_{B(y,R)^{\complement}} < \tau_A] = \frac{a(x) - \mathbb{E}_x a(S_{\tau_A})}{\mathbb{E}_x(a(S_{\tau_{B(y,R)^{\complement}}}) - a(S_{\tau_A}) \mid \tau_{B(y,R)^{\complement}} < \tau_A)} \qquad (3.51)$$

(using (3.39))

$$= \frac{q_A(x)}{a(R) - O(\ln b) + O(\frac{\|y\|+1}{R})}, \qquad (3.52)$$

and this implies (3.50). $\qquad\qquad\square$

Since the limit in (3.50) does not depend on y, this means that Proposition 3.14 indeed shows that the definition (3.49) of q_A does not depend on the choice of $y_0 \in A$ (since "moving y_0 within A" effectively amounts to changing y in (3.50)).

Now we are ready to define the notion of the harmonic measure in two dimensions:

Definition 3.15. For a finite set $A \subset \mathbb{Z}^2$, the *harmonic measure* ($\text{hm}_A(y), y \in A$) is defined as follows:

$$\text{hm}_A(y) = \frac{1}{4} \sum_{z \sim y} q_A(z) = \frac{1}{4} \sum_{\substack{z \notin A: \\ z \sim y}} (a(z) - \mathbb{E}_z a(S_{\tau_A})). \qquad (3.53)$$

Admittedly, at this point it may be not completely clear why $\text{hm}_A(\cdot)$ should be even nonnegative (it is because of Proposition 3.14); what is definitely not clear, is why it sums to 1 on ∂A. Things start to make sense, though, when we observe that (3.53) is very similar to (3.20) (i.e., the corresponding definition in the many-dimensional case): the harmonic measure is proportional to the escape probability. To see this, observe that Proposition 3.14 implies that

$$\text{hm}_A(y) = \lim_{R\to\infty} a(R)\mathbb{P}_y[\tau_A^+ > \tau_{B(R)^C}] = \frac{2}{\pi}\lim_{R\to\infty}\mathbb{P}_y[\tau_A^+ > \tau_{B(R)^C}]\ln R; \quad (3.54)$$

indeed, just write, conditioning on the first step,

$$\mathbb{P}_y[\tau_A^+ > \tau_{B(R)^C}] = \frac{1}{4}\sum_{\substack{z\notin A:\\ z\sim y}}\mathbb{P}_z[\tau_{B(R)^C} < \tau_A],$$

then multiply both sides by $a(R)$ and pass to the limit.

To complete the analogy with the many-dimensional case, we have to prove that the harmonic measure is the entrance law "starting at infinity" (compare to Theorem 3.8; since the walk is recurrent in two dimensions, we do not need to condition on $\{\tau_A < \infty\}$):

Theorem 3.16. *For all finite $A \subset \mathbb{Z}^2$ and all $y \in A$, we have*

$$\text{hm}_A(y) = \lim_{x\to\infty}\mathbb{P}_x[S_{\tau_A} = y]. \quad (3.55)$$

Proof First, it is clear that, without restricting generality, we can assume that $0 \in A$. Recall the notation $N_x^b = \sum_{k=0}^{\tau_A^+-1}\mathbf{1}\{S_k = x\}$ from the proof of Theorem 3.8; also, let us define

$$N_{x,R}^b = \sum_{k=0}^{(\tau_A^+-1)\wedge\tau_{B(R)^C}}\mathbf{1}\{S_k = x\}, \qquad N_{x,R}^\# = \sum_{k=\tau_A^+}^{\tau_{B(R)^C}}\mathbf{1}\{S_k = x\},$$

$$N_{x,R} := N_{x,R}^b + N_{x,R}^\# = \sum_{k=0}^{\tau_{B(R)^C}}\mathbf{1}\{S_k = x\}$$

to be the corresponding (i.e., before the first re-entry, after the first re-entry, and total) visit counts "restricted" on $B(R)$. Quite similarly to the proof of Theorem 3.8, one can write

$$\mathbb{P}_x[S_{\tau_A} = y]$$
$$= \mathbb{E}_y N_x^b$$

(by the Monotone Convergence Theorem)

$$= \lim_{R\to\infty}\mathbb{E}_y N_{x,R}^b$$
$$= \lim_{R\to\infty}(\mathbb{E}_y N_{x,R} - \mathbb{E}_y N_{x,R}^\#)$$
$$= \lim_{R\to\infty}\left(G_{B(R)}(y,x) - \sum_{z\in A}\mathbb{P}_y[\tau_A^+ < \tau_{B(R)^C}, S_{\tau_A^+} = z]G_{B(R)}(z,x)\right)$$

(using (3.47))

$$= \lim_{R \to \infty} \left(a(R) - a(y - x) + O(\tfrac{\|y\|+1}{R}) \right.$$

$$\left. - \sum_{z \in A} \mathbb{P}_y[\tau_A^+ < \tau_{B(R)^C}, S_{\tau_A^+} = z](a(R) - a(z - x) + O(\tfrac{\|z\|+1}{R})) \right)$$

(note that $\mathbb{P}_y[\tau_A^+ < \tau_{B(R)^C}, S_{\tau_A^+} = z] \to \mathbb{P}_y[S_{\tau_A^+} = z]$ as $R \to \infty$)

$$= \lim_{R \to \infty} a(R) \left(1 - \sum_{z \in A} \mathbb{P}_y[\tau_A^+ < \tau_{B(R)^C}, S_{\tau_A^+} = z] \right)$$

$$- a(y - x) + \sum_{z \in A} \mathbb{P}_y[S_{\tau_A^+} = z]a(z - x)$$

(observe that, in the first parentheses, we have $\mathbb{P}_y[\tau_A^+ > \tau_{B(R)^C}]$, then use (3.54))

$$= \mathrm{hm}_A(y) - a(y - x) + \sum_{z \in A} \mathbb{P}_y[S_{\tau_A^+} = z]a(z - x). \tag{3.56}$$

From (3.38), it is straightforward to obtain that

$$a(y - x) - \sum_{z \in A} \mathbb{P}_y[S_{\tau_A^+} = z]a(z - x)$$

$$= \sum_{z \in A} \mathbb{P}_y[S_{\tau_A^+} = z](a(y - x) - a(z - x))$$

$$= O(\tfrac{\mathrm{diam}(A)}{\|x\|}),$$

which converges to 0 as $x \to \infty$, and so the proof of (3.55) is concluded. $\square$

Then, Theorem 3.16 implies that $\mathrm{hm}_A(\cdot)$ is indeed a probability measure (because, due to the recurrence, so is the entrance measure to A for any fixed starting point).

We are happy with Theorem 3.16, but not completely so. This is because it is a *qualitative* result, which does not say how fast the convergence occurs. Imagine, for example, that the set A is "large" (say, a disk of radius r^2), and the distance from x to A is even larger (say, of order r^3). How does the entrance measure from x to A compare to the harmonic measure then? Well, in the end of the proof of Theorem 3.16 we obtained *some* estimate, but it is not quite sharp – in the example we just considered, the term $O(\tfrac{\mathrm{diam}(A)}{\|x\|})$ would be of order r^{-1}, while $\mathrm{hm}_A(y)$ itself would be of order r^{-2} (since there are $O(r^2)$ sites on the boundary of the disk), which, to put it mildly, is not quite satisfactory. The following theorem gives a much better estimate:

Theorem 3.17. *Let A be a finite subset of* $\mathbb{Z}^2$ *and assume that* $\mathrm{dist}(x, A) \geq 3\,\mathrm{diam}(A) + 1$. *Then it holds that*

$$\mathbb{P}_x[S_{\tau_A} = y] = \mathrm{hm}_A(y)(1 + O(\tfrac{\mathrm{diam}(A)}{\mathrm{dist}(x,A)})). \tag{3.57}$$

Proof Again, without restricting generality, we assume that $0 \in A$ and $|A| \geq 2$, so $\mathrm{diam}(A) \geq 1$. Recall that in the calculation (3.56) we obtained

$$\mathbb{P}_x[S_{\tau_A} = y] - \mathrm{hm}_A(y) = -a(y - x) + \sum_{z \in A} \mathbb{P}_y[S_{\tau_A^+} = z]a(z - x). \tag{3.58}$$

The idea is to estimate the right-hand side of the preceding expression using a martingale argument similar to that in the proof of Proposition 3.14. We need some preparations, though. From the asymptotic expression (3.36) for a, it is straightforward to obtain that there exist constants $\theta_{1,2} > 0$ such that whenever $\|x\| > \theta_1$ and $2\|y\| \leq \|x\|$ it holds that $a(x) - a(y) > \theta_2$ (in fact, it is even clear that for any $\theta_2 < \frac{2}{\pi}\ln 2$ it is possible to choose a large enough θ_1 such that the preceding holds, but we do not need to be so precise).

Let us now abbreviate $V = \partial_e B((2\,\mathrm{diam}(A)) \vee \theta_1)$, so that, due to the preceding discussion,

$$a(v) - a(z) \geq \theta_2 \qquad \text{for all } v \in V \text{ and } z \in A. \tag{3.59}$$

We assume additionally that $\|x\| > \theta_1$; the reader is invited to check (or to just accept) that this assumption does not restrict generality.

Let us apply the optional stopping theorem to the martingale $a(S_{n \wedge \tau_x} - x)$ and the stopping time $\tau_A^+ \wedge \tau_V$ (see Figure 3.6; observe that $\tau_V < \tau_x$ for the walk that starts at $y \in \partial A$)

$$a(y - x) = \mathbb{E}_y a(S_{\tau_A^+ \wedge \tau_V} - x)$$
$$= \mathbb{E}_y(a(S_{\tau_A^+} - x)\mathbf{1}\{\tau_A^+ < \tau_V\}) + \mathbb{E}_y(a(S_{\tau_V} - x)\mathbf{1}\{\tau_V < \tau_A^+\})$$

(writing $\mathbf{1}\{\tau_A^+ < \tau_V\} = 1 - \mathbf{1}\{\tau_V < \tau_A^+\}$ in the first term)

$$= \mathbb{E}_y a(S_{\tau_A^+} - x)$$
$$\quad + \mathbb{E}_y((a(S_{\tau_V} - x) - a(S_{\tau_A^+} - x))\mathbf{1}\{\tau_V < \tau_A^+\})$$
$$= \sum_{z \in A} \mathbb{P}_y[S_{\tau_A^+} = z]a(z - x)$$
$$\quad + \mathbb{E}_y(a(S_{\tau_V} - x) - a(S_{\tau_A^+} - x) \mid \tau_V < \tau_A^+)\mathbb{P}_y[\tau_V < \tau_A^+].$$

So, the preceding together with (3.58) implies that

$$\mathrm{hm}_A(y) - \mathbb{P}_x[S_{\tau_A} = y] = \mathbb{E}_y(a(S_{\tau_V} - x) - a(S_{\tau_A^+} - x) \mid \tau_V < \tau_A^+)$$
$$\times \mathbb{P}_y[\tau_V < \tau_A^+]. \tag{3.60}$$

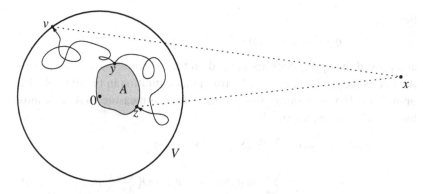

Figure 3.6 On the proof of Theorem 3.17: the random walk starts at $y \in \partial A$ and ends either on the first re-entry to A or entry to V.

Let us now recall the expression (3.51) from the proof of Proposition 3.14 together with the definition (3.53) of the harmonic measure. For the second factor in the right-hand side of (3.60), we have

$$
\mathbb{P}_y[\tau_V < \tau_A^+] = \frac{1}{4} \sum_{\substack{z \notin A \\ z \sim y}} \mathbb{P}_u[\tau_V < \tau_A^+]
$$

$$
= \frac{1}{4} \sum_{\substack{z \notin A \\ z \sim y}} \frac{q_A(z)}{\mathbb{E}_z(a(S_{\tau_V}) - a(S_{\tau_A}) \mid \tau_V < \tau_A)}
$$

(by our choice of V and (3.59))

$$
\leq \frac{\mathrm{hm}_A(y)}{\theta_2}.
$$

We also to observe that, for any $v \in V$ and $z \in A$, from (3.38) we obtain that $a(x - v) - a(x - z) = O(\frac{\mathrm{diam}(A)}{\mathrm{dist}(x,A)})$ (look again at Figure 3.6), so the first factor in the right-hand side of (3.60) is $O(\frac{\mathrm{diam}(A)}{\mathrm{dist}(x,A)})$ as well. This shows that the right-hand side of (3.60) is indeed $O(\frac{\mathrm{diam}(A)}{\mathrm{dist}(x,A)}) \times \mathrm{hm}_A(y)$ and therefore concludes the proof of Theorem 3.17. □

Capacity

When one learns something new, it is a good idea to review one's old notes and see if there was something interesting that went unnoticed at that time. Specifically, let us revisit the calculations in and around Proposition 3.14. Recall that, for a finite $A \subset \mathbb{Z}^2$ and $y_0 \in A$, we defined in (3.49) the quantities

$$q_A(x) = a(x - y_0) - \mathbb{E}_x a(S_{\tau_A} - y_0), \qquad x \in \mathbb{Z}^2,$$

and proved that $q_A(x)$ does not depend on the choice of y_0 (as long as $y_0 \in A$). Let us reexamine the second term in that definition in the light of Theorem 3.17. If x is far away from A, the entrance measure to A is "almost harmonic" and, quantitatively,

$$\mathbb{E}_x a(S_{\tau_A} - y_0) = \sum_{z \in A} \mathbb{P}_x[S_{\tau_A} = z] a(z - y_0)$$

$$= \sum_{z \in A} \mathrm{hm}_A(z) a(z - y_0)\left(1 + O\left(\tfrac{\mathrm{diam}(A)}{\mathrm{dist}(x,A)}\right)\right). \tag{3.61}$$

The main term in (3.61) does not depend on x and therefore looks as something important. It is so indeed:

Definition 3.18. For a finite set A with $y_0 \in A$, we define its *capacity* by

$$\mathrm{cap}(A) = \sum_{x \in A} a(x - y_0)\,\mathrm{hm}_A(x). \tag{3.62}$$

First, we need to show that Definition 3.18 does not depend on the choice of $y_0 \in A$. Basically, this is a consequence of the fact that $q_A(x)$ does not depend on the choice of $y_0 \in A$. Indeed, if $y_1 \in A$, we also have $q_A(x) = a(x - y_1) - \mathbb{E}_x a(S_{\tau_A} - y_1)$ and so, by (3.61),

$$a(x - y_0) - a(x - y_1) = \left(\mathrm{cap}(A) - \sum_{z \in A} \mathrm{hm}_A(z) a(z - y_1)\right)\left(1 + O\left(\tfrac{\mathrm{diam}(A)}{\mathrm{dist}(x,A)}\right)\right).$$

Since the left-hand side of the preceding equation clearly converges to 0 as $x \to \infty$, the expression in the first parentheses in the right-hand side must be *equal* to 0 (since it does not depend on x).

Now, assume that $y_0 \in A \subset B(r_A)$. Then (3.61) implies that

$$q_A(x) = a(x - y_0) - \mathrm{cap}(A)\left(1 + O\left(\tfrac{r_A}{\|x\|}\right)\right),$$

and then, recalling the calculation (3.51), we can write[21]

$$\mathbb{P}_x[\tau_{B(R)^\complement} < \tau_A] = \frac{a(x - y_0) - \mathrm{cap}(A)\left(1 + O\left(\tfrac{r_A}{\|x\|}\right)\right)}{a(R) + O(R^{-1}) - \mathrm{cap}(A)\left(1 + O\left(\tfrac{r_A}{\|x\|}\right)\right)}. \tag{3.63}$$

That is, if we know the capacity of A, we are then able to compute the escape probabilities as before with higher precision. Notice that, the larger

---

[21] Observe that $\mathbb{E}_x(S_{\tau_A} \mid \tau_{B(R)^\complement} < \tau_A)$ is also *quite* close to $\mathrm{cap}(A)$, as can be seen by conditioning on the location of $S_{\tau_{B(R)^\complement}}$.

cap(A) is, the smaller is the probability in (3.63). This again justifies the intuition "the capacity measures how big is the set from the point of view of SRW"; only in two dimensions it enters to the second-order term (and not to the principal one, as in the higher-dimensional case).

Before going further, let us also obtain a finer expression on the escape probability from $y \in \partial A$ to the boundary of a large disk. Assume that $0 \in A$; then, analogously to the previous calculation, we can write

$$\mathbb{P}_y[\tau_A^+ > \tau_{\mathsf{B}(R)^\complement}] = \frac{1}{4} \sum_{\substack{z \notin A \\ z \sim y}} \mathbb{P}_z[\tau_A^+ > \tau_{\mathsf{B}(R)^\complement}]$$

$$= \frac{1}{4} \sum_{\substack{z \notin A \\ z \sim y}} \frac{q_A(z)}{a(R) + O(R^{-1}) - \mathrm{cap}(A)(1 + O(\frac{r_A}{\|x\|}))}$$

$$= \frac{\mathrm{hm}_A(z)}{a(R) + O(R^{-1}) - \mathrm{cap}(A)(1 + O(\frac{r_A}{\|x\|}))}. \tag{3.64}$$

Now, let us discuss the simplest cases where the two-dimensional capacities can be calculated. To start with, what can we say about capacities of one- and two-point sets? The first question is easy: since $a(0) = 0$, it holds that $\mathrm{cap}(\{x\}) = 0$ for any $x \in \mathbb{Z}^2$. As for two-point sets, observe that, by symmetry, the harmonic measure of any two-point set is uniform, so

$$\mathrm{cap}(\{x, y\}) = \frac{a(y - x)}{2} \tag{3.65}$$

for any $x, y \in \mathbb{Z}^2$, $x \ne y$. As for the capacity of a disk, (3.39) implies that

$$\mathrm{cap}(\mathsf{B}(r)) = a(r) + O(r^{-1}). \tag{3.66}$$

It is remarkable to observe that the capacities of a two-point set $\{0, x\}$ with $\|x\| = r$ and the whole disk $\mathsf{B}(r)$ only differ by a factor of 2 (this is asymptotically, as $r \to \infty$). Dimension two sometimes brings surprises.

Next, it is not difficult to obtain from (3.63) that

$$\mathrm{cap}(A) = \lim_{x \to \infty} \left(a(x - y_0) - \lim_{R \to \infty} (a(R)\mathbb{P}_x[\tau_{\mathsf{B}(R)^\complement} < \tau_A]) \right) \tag{3.67}$$

(first, multiply it by $a(R)$ and let $R \to \infty$ to get rid of the denominator, then "separate" the term cap(A) from the numerator). An easy corollary of this fact is the following:

Proposition 3.19. *(i) Assume that $A \subset B$ are finite nonempty subsets of $\mathbb{Z}^2$. Then* $\mathrm{cap}(A) \le \mathrm{cap}(B)$.
(ii) Let A, B be finite subsets of $\mathbb{Z}^2$ such that $A \cap B$ is nonempty. Then $\mathrm{cap}(A \cup B) \le \mathrm{cap}(A) + \mathrm{cap}(B) - \mathrm{cap}(A \cap B)$.

Proof Note that, in both cases, we may choose $y_0 \in A \cap B$. The part (i) is straightforward: just observe that $\mathbb{P}_x[\tau_{\mathsf{B}(R)^\complement} < \tau_A] \geq \mathbb{P}_x[\tau_{\mathsf{B}(R)^\complement} < \tau_B]$ for $A \subset B$ (indeed, it is easier to avoid a smaller set), and use (3.67). The proof of part (ii) is only a bit more complicated: write

$$\mathbb{P}_x[\tau_{\mathsf{B}(R)^\complement} < \tau_{A \cup B}]$$
$$= 1 - \mathbb{P}_x[\{\tau_A < \tau_{\mathsf{B}(R)^\complement}\} \cup \{\tau_B < \tau_{\mathsf{B}(R)^\complement}\}]$$
$$= 1 - \mathbb{P}_x[\tau_A < \tau_{\mathsf{B}(R)^\complement}] - \mathbb{P}_x[\tau_B < \tau_{\mathsf{B}(R)^\complement}] + \mathbb{P}_x[\tau_A \vee \tau_B < \tau_{\mathsf{B}(R)^\complement}]$$

(since $\tau_{A \cap B} \geq \tau_A \vee \tau_B$)

$$\geq 1 - \mathbb{P}_x[\tau_A < \tau_{\mathsf{B}(R)^\complement}] - \mathbb{P}_x[\tau_B < \tau_{\mathsf{B}(R)^\complement}] + \mathbb{P}_x[\tau_{A \cap B} < \tau_{\mathsf{B}(R)^\complement}]$$
$$= \mathbb{P}_x[\tau_{\mathsf{B}(R)^\complement} < \tau_A] + \mathbb{P}_x[\tau_{\mathsf{B}(R)^\complement} < \tau_B] - \mathbb{P}_x[\tau_{\mathsf{B}(R)^\complement} < \tau_{A \cap B}],$$

and then use (3.67) again. □

Observe that is it essential to require in (ii) that these two sets have nonempty intersection – otherwise $A = \{x\}$ and $B = \{y\}$ with $x \neq y$ would, sort of, provide a counterexample. A possible way to get around it (so that (ii) would be valid for *all* A, B) would be to declare the capacity of an empty set to be equal to $(-\infty)$ – this is formally in agreement with (3.67), by the way.

To conclude this section, we formulate and prove a result analogous to Theorem 3.17 (which is, in fact, its corollary; so we let the proof be rather sketchy), but for the *conditional* entrance measure. This result will be important for applications, since it is frequent that one needs to consider entrance measures which are not "clean" (i.e., they are conditioned on something else in addition).

Theorem 3.20. *Let A be a finite subset of $\mathbb{Z}^2$ and assume that* $\text{dist}(x, A) \geq 5 \, \text{diam}(A) + 1$. *Assume additionally that (finite or infinite) $A' \subset \mathbb{Z}^2$ is such that* $\text{dist}(A, A') \geq \text{dist}(x, A) + 1$. *Then it holds that*

$$\mathbb{P}_x[S_{\tau_A} = y \mid \tau_A < \tau_{A'}] = \text{hm}_A(y)(1 + O(\tfrac{\Psi \, \text{diam}(A)}{\text{dist}(x,A)})), \qquad (3.68)$$

where $\Psi = (a(\text{dist}(x, A)) - \text{cap}(A)) \vee 1$.

Note that, in the case $A = \mathsf{B}(r)$ and $\|x\| = s$, it holds that $O(\tfrac{\Psi \, \text{diam}(A)}{\text{dist}(x,A)}) = O(\tfrac{r}{s} \ln \tfrac{s}{r})$, which is the usual error term that one finds in the literature (see e.g. lemma 2.2 of [34]).

Proof Let us assume without restricting generality that $0 \in A$, $A \subset \mathsf{B}(r)$ for $r \leq \text{diam}(A)$, and abbreviate $\|x\| = s$. First, using the usual last-exit-decomposition reasoning, it is clear that it is enough to prove (3.68) for

$A' = B(s)^C$. Then, since on its way to A the walker must pass through $\partial B(\frac{2}{3}s)$, it would be enough to prove that

$$\mathbb{P}_z[S_{\tau_A} = y \mid \tau_A < \tau_{B(s)^C}] = hm_A(y)(1 + O(\tfrac{\Psi r}{s})) \qquad (3.69)$$

for any $z \in \partial B(\frac{2}{3}s)$. We write

$$\mathbb{P}_z[S_{\tau_A} = y, \tau_A < \tau_{B(s)^C}]$$
$$= \mathbb{P}_z[S_{\tau_A} = y] - \mathbb{P}_z[S_{\tau_A} = y, \tau_{B(s)^C} < \tau_A]$$
$$= \mathbb{P}_z[S_{\tau_A} = y] - \sum_{z' \in B(s)^C} \mathbb{P}_z[S_{\tau_{B(s)^C}} = z', \tau_{B(s)^C} < \tau_A]\mathbb{P}_{z'}[S_{\tau_A} = y]$$

(by Theorem 3.17)

$$= hm_A(y)(1 + O(\tfrac{r}{s})) - \mathbb{P}_z[\tau_{B(s)^C} < \tau_A] \, hm_A(y)(1 + O(\tfrac{r}{s}))$$
$$= hm_A(y)(\mathbb{P}_z[\tau_A < \tau_{B(s)^C}] + O(\tfrac{r}{s}))$$

and, since (3.64) implies that $\mathbb{P}_z[\tau_A < \tau_{B(s)^C}] \geq O(1/\Psi)$, we obtain (3.69) by dividing the preceding by $\mathbb{P}_z[\tau_A < \tau_{B(s)^C}]$. $\qquad \square$

3.3 Exercises

Exercise 3.1. Prove that in Proposition 3.1 it is enough to assume that h is bounded only from one side (i.e., either $h(x) \leq K$ for all x, or $h(x) \geq K$ for all x).

Transient case (Section 3.1)

Exercise 3.2. Prove that $G(0) - G(e_1) = 1$.

Exercise 3.3. Prove the following *exact* expression for Green's function:

$$G(x) = \frac{1}{(2\pi)^d} \int_{[-\pi,\pi]^d} \frac{e^{i(\theta,x)}}{1 - \Phi(\theta)} \, d\theta, \qquad (3.70)$$

where $\Phi(\theta) = d^{-1} \sum_{k=1}^d \cos \theta_k$. In particular, it holds that

$$G(0) = \frac{1}{(2\pi)^d} \int_{[-\pi,\pi]^d} \frac{1}{1 - \Phi(\theta)} \, d\theta. \qquad (3.71)$$

Exercise 3.4. Follow the proof of theorem 4.3.1 of [63] to see how to obtain (3.5) from the local CLT.

Exercise 3.5. Assume that we *only* know that, for some $c > 0$, $G(x) \sim c\|x\|^{-(d-2)}$ as $x \to \infty$. Prove that $\frac{d-2}{2c}$ equals the volume of the unit ball in $\mathbb{R}^d$.

Exercise 3.6. Obtain a direct proof (i.e., a proof that does not use (3.5)) that $\mathrm{Es}_{\mathsf{B}(r)}(y) \asymp r^{-1}$ for any $y \in \mathsf{B}(R)^{\complement}$ (and therefore also that $\mathrm{cap}(\mathsf{B}(r)) \asymp r^{d-2}$). Suggestion: use Lyapunov functions.

Exercise 3.7. Note that the definition (3.43) of the restricted Green's function makes sense in any dimension. For $d \geq 3$, prove an analogue of Theorem 3.13's statement:

$$G_\Lambda(x, y) = G(x - y) - \mathbb{E}_x G(S_{\tau_{\Lambda^\complement}} - y). \tag{3.72}$$

Exercise 3.8. For $d \geq 2$ and $r \geq 1$, prove that there exist $c_2 > c_1 > 0$ (depending only on dimension) such that

$$\frac{c_1}{r^{d-1}} \leq \mathbb{P}_x[S_{\tau_{\partial_e\mathsf{B}(r)}} = y] \leq \frac{c_2}{r^{d-1}} \tag{3.73}$$

for all $x \in \mathsf{B}(r/4)$ and all $y \in \partial_e\mathsf{B}(r)$.

Exercise 3.9. For $A \subset \mathbb{Z}^d$ and $x \in \mathbb{Z}^d$, denote $G(x, A) = \sum_{y \in A} G(x, y)$ to be the mean number of visits to A starting from x. Prove that, for finite A,

$$\frac{|A|}{\max_{y \in A} G(y, A)} \leq \mathrm{cap}(A) \leq \frac{|A|}{\min_{y \in A} G(y, A)}. \tag{3.74}$$

We have to mention that (3.74) provides another useful way of estimating capacities.

Exercise 3.10. For a finite $A \subset \mathbb{Z}^d$, let $\mathcal{K}_A$ be a class of nonnegative functions, defined in the following way:

$$\mathcal{K}_A = \Big\{h: \mathbb{Z}^d \to \mathbb{R}_+ \text{ such that } h(x) = 0 \text{ for all } x \notin A$$
$$\text{and } \sum_{x \in \mathbb{Z}^d} G(y, x)h(x) \leq 1 \text{ for all } y \in \mathbb{Z}^d\Big\}.$$

Prove that

$$\mathrm{cap}(A) = \sup_{h \in \mathcal{K}_A} \sum_{x \in A} h(x).$$

Exercise 3.11. Prove that the capacity (defined as in (3.10)) of any infinite transient set is infinite.

Exercise 3.12. *Estimate* (in the sense of "$\asymp$") the capacities of various sets (and in various dimensions), such as

- a line segment (i.e., a sequence of neighbouring sites lying on a straight line);
- a "plaquette" (i.e., a discrete two-dimensional square immersed in $\mathbb{Z}^d$);

- a cylinder (product of a line segment with a $(d-1)$-dimensional plaquette), with hight/width ratio varying arbitrarily;
- whatever else you can imagine.

Exercise 3.13. Prove the "only if" part of Theorem 3.7.

Exercise 3.14. Prove the "if" part of Theorem 3.7.

Exercise 3.15. Using Wiener's criterion, prove that, in three dimensions, the "ray" $\{(0,0,k), k \geq 0\} \subset \mathbb{Z}^3$ is a recurrent set.

Exercise 3.16. Give an example of a transient set such that the *expected* number of visits there is infinite.

Exercise 3.17. Let $f: \mathbb{R}_+ \to (0,1]$ be a monotonously decreasing function with $\lim_{s \to \infty} f(s) = 0$. Let us construct a *random* set $A_f \subset \mathbb{Z}^d$ in the following way: $x \in \mathbb{Z}^d$ is included to A_f with probability $f(\|x\|)$, independently.

 (i) Is it true that, for any such f, $\mathbb{P}[A_f$ is transient] must be either 0 or 1?
 (ii) Give (nontrivial) examples of functions for which the resulting random set is recurrent/transient. What should be the "critical rate of decay" of f which separates recurrence from transience?

Exercise 3.18. Give an example of a transient set $A \subset \mathbb{Z}^d$ such that $\mathbb{P}_x[\tau_A < \infty] = 1$ for infinitely many $x \in \mathbb{Z}^d$.

Exercise 3.19. For any transient set A and any $\varepsilon > 0$, prove that $\mathbb{P}_x[\tau_A < \infty] < \varepsilon$ for infinitely many $x \in \mathbb{Z}^d$.

Exercise 3.20. Prove that, for finite A,

$$\sum_{y \in A} \mathrm{hm}_A(y) G(y, A) = \frac{|A|}{\mathrm{cap}(A)}. \tag{3.75}$$

Exercise 3.21. For $r \geq 1$, prove that $\mathrm{hm}_{B(r)}(x) \asymp r^{-(d-1)}$ for all $x \in \partial B(r)$.

Exercise 3.22. Show that, for finite $A \subset B \subset \mathbb{Z}^d$ $(d \geq 3)$. it holds

$$\mathbb{P}_{\mathrm{hm}_B}[\tau_A < \infty] = \frac{\mathrm{cap}(A)}{\mathrm{cap}(B)} \tag{3.76}$$

(here $\mathbb{P}_{\mathrm{hm}_B}$ means the probability for the walk with the initial position chosen according to the measure hm_B).

Exercise 3.23. Prove that the harmonic measure is *consistent*, in the sense that, for any finite $A \subset B$,

$$\mathbb{P}_{\mathrm{hm}_B}[S_{\tau_A} = y \mid \tau_A < \infty] = \mathrm{hm}_A(y) \tag{3.77}$$

for all $y \in A$.

Exercise 3.24. Prove that for *any* finite $A, B \subset \mathbb{Z}^d, d \geq 3$, it holds that

$$\mathbb{P}_{\mathrm{hm}_A}[\tau_B < \infty] \leq \frac{\mathrm{cap}(B)}{\mathrm{cap}(A)}. \tag{3.78}$$

Exercise 3.25. Can you obtain an analogue of Theorem 3.17 in the many-dimensional case (i.e., $d \geq 3$)? That is, prove that

$$\mathbb{P}_x[S_{\tau_A} = y \mid \tau_A < \infty] = \mathrm{hm}_A(y)(1 + O(\tfrac{\mathrm{diam}(A)}{\mathrm{dist}(x,A)})). \tag{3.79}$$

Exercise 3.26. Let $A \subset \mathbb{Z}^d$ be finite and $x \notin A, y \in \partial A$.

(i) Prove that

$$\mathbb{P}_x[\tau_A < \infty, S_{\tau_A} = y] = \mathbb{P}_y[\tau_x < \tau_A^+](G(0) - O(\tfrac{(\mathrm{diam}(A))^{d-2}}{(\mathrm{dist}(x,A))^{2d-4}})), \tag{3.80}$$

and that $\mathbb{P}_x[\tau_A < \infty, S_{\tau_A} = y] > G(0)\mathbb{P}_y[\tau_x < \tau_A^+]$ (that is, the $O(\cdot)$ in (3.80) is always strictly positive).

(ii) Prove that

$$\mathbb{P}_y[\tau_x < \infty, \tau_A^+ = \infty] = \mathbb{P}_y[\tau_x < \infty] \, \mathrm{Es}_A(y)(1 + O(\tfrac{\mathrm{diam}(A)}{\mathrm{dist}(x,A)})), \tag{3.81}$$

that is, the events {eventually hit x} and {escape from A} are approximately independent under $\mathbb{P}_y$ when $\frac{\mathrm{diam}(A)}{\mathrm{dist}(x,A)} \to 0$.

Exercise 3.27. Consider a transient and *reversible* Markov chain with the reversible measure μ. Prove that (for Green's function G defined in the usual way)

$$\mu(x)G(x, y) = \mu(y)G(y, x)$$

for any x, y. Prove that the preceding also holds for the *restricted* Green's function G_Λ.

Recurrent case (Section 3.2)

Exercise 3.28. Prove, preferably without any calculations, that

(i) $\mathbb{P}_0[S_{2n} = 0] > \mathbb{P}_0[S_{2n} = e_1 + e_2]$ for all n.
(ii) $\mathbb{P}_0[S_k = x] \geq \mathbb{P}_0[S_k = x + 2e_1]$, for all k and x such that $x \cdot e_1 \geq 0$, and the inequality is strict in the case when the first probability is strictly positive.
(iii) Use the preceding to conclude that, for any $x \neq 0$ and any n,

$$\mathbb{P}_0[S_{2n} = 0] > \mathbb{P}_0[S_{2n} = x] + \mathbb{P}_0[S_{2n+1} = x]. \tag{3.82}$$

Exercise 3.29. Can you simplify the proof of existence of the potential kernel a using the coordinate decoupling idea from the end of Section 2.1?

Exercise 3.30. Give a rigorous proof that for any d and any *finite* $\Lambda \subset \mathbb{Z}^d$ it holds that $\mathbb{E}_x \tau_\Lambda c < \infty$ for all x.

Exercise 3.31. Prove that

$$a(x) = \frac{1}{(2\pi)^2} \int_{[-\pi,\pi]^2} \frac{1 - \cos(\theta_1 x_1 + \theta_2 x_2)}{1 - \frac{1}{2}(\cos\theta_1 + \cos\theta_2)} \, d\theta. \qquad (3.83)$$

Exercise 3.32. Derive (3.36) either by analysing (3.83) (cf. section III.12 of [91]) or by analysing (3.25) directly, via the Local Central Limit Theorem (cf. section 4.4 of [63]).

Exercise 3.33. In the proof of Theorem 3.13, show that

$$a(x - y) = \mathbb{E}_x Y_{\tau_\Lambda c}$$

without invoking the optional stopping theorem.

Exercise 3.34. Let $x \neq 0$ and define

$$\eta_x = \sum_{k=0}^{\tau_0^+ - 1} \mathbf{1}\{S_k = x\}$$

to be the number of visits to x before hitting the origin. Prove that $\mathbb{E}_x \eta_x = 2a(x)$ and $\mathbb{E}_0 \eta_x = 1$.

Exercise 3.35. Prove that the harmonic measure is consistent in two dimensions as well (recall Exercise 3.23): for any finite $A \subset B$,

$$\mathbb{P}_{\mathrm{hm}_B}[S_{\tau_A} = y] = \mathrm{hm}_A(y) \qquad (3.84)$$

for all $y \in A$.

Exercise 3.36. Show that, in general, it is not possible to get rid of logarithms in the error terms in (3.68) and (3.69) (i.e., substitute $O(\frac{\Psi r}{s})$ by $O(\frac{r}{s})$ there; recall that the term Ψ is logarithmic) by considering the following example. For $k \geq 5$, let $\Lambda = [-k, k]^2 \subset \mathbb{Z}^2$ be the (discrete) square of size $2k$ centred at the origin, let $y = e_1, z = -e_1$, and let $x \in \Lambda$ be such that $x \cdot e_1 \in [k/3, 2k/3]$ and $x \cdot e_2 \in [-k/2, k/2]$ (see Figure 3.7).

(i) Show that $\mathbb{P}_x[\tau_{y,z} < \tau_\Lambda c] \asymp \frac{1}{\ln k}$.

(ii) Prove that

$$\mathbb{P}_x[\tau_y < \tau_\Lambda c, S_j \cdot e_1 > 0 \text{ for all } j \leq \tau_y] \asymp \frac{1}{k}.$$

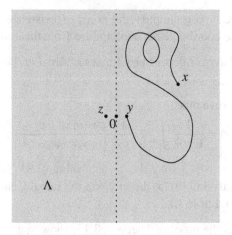

Figure 3.7 The setup of Exercise 3.36.

(iii) From (i) and (ii), conclude that

$$\mathbb{P}_x[S_{\tau_{y,z}} = y \mid \tau_{y,z} < \tau_{\Lambda^C}] - \mathbb{P}_x[S_{\tau_{y,z}} = z \mid \tau_{y,z} < \tau_{\Lambda^C}] \asymp \frac{\ln k}{k}. \quad (3.85)$$

Do you have a heuristic explanation about why there is this difference from the unconditional entrance measure result (Theorem 3.17), where we do not have logarithmic terms?

Exercise 3.37. Prove that if $x \in A \subset B$, then $hm_A(x) \geq hm_B(x)$.

Exercise 3.38. Prove that

$$\text{cap}(A) = \left(\sup_{y \in A} f(y) \right)^{-1},$$

where the supremum is over all nonnegative functions f on A such that $\sum_{y \in A} a(x - y) f(y) \leq 1$ for all $x \in A$.

Exercise 3.39. Let $x_1, x_2, x_3 \in \mathbb{Z}^2$ be three *distinct* sites, and abbreviate $v_1 = x_2 - x_1, v_2 = x_3 - x_2, v_3 = x_1 - x_3$. Prove that the capacity of the set $A = \{x_1, x_2, x_3\}$ is given by the formula

$$\frac{a(v_1)a(v_2)a(v_3)}{a(v_1)a(v_2) + a(v_1)a(v_3) + a(v_2)a(v_3) - \frac{1}{2}(a^2(v_1) + a^2(v_2) + a^2(v_3))}. \quad (3.86)$$

Exercise 3.40. What if we try to develop the same theory as in Section 3.1, only on a finite set $\Lambda \subset \mathbb{Z}^2$ instead of the whole space $\mathbb{Z}^d$, $d \geq 3$? The

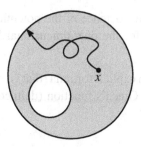

Figure 3.8 The walk hits the outer boundary.

analogy is clear: we regard the outside of Λ as "infinity". Let $A \subset \Lambda$. Analogously to (3.9), let us define for any $x \in \Lambda$

$$\mathrm{Es}_{A,\Lambda}(x) = \mathbb{P}_x[\tau_A^+ > \tau_{\Lambda^c}]\mathbf{1}\{x \in A\};$$

on A, this equals the probability of "escaping from A within Λ". Also, let

$$u_{A,\Lambda}(x) = \mathbb{P}_x[\tau_A < \tau_{\Lambda^c}]$$

be the probability of "hitting A within Λ"; note that $u_{A,\Lambda} = 1$ for all $x \in A$.

 (i) Prove a relation analogous to (3.11):

$$u_{A,\Lambda}(y) = \sum_{x\in\Lambda} G_\Lambda(y, x)\,\mathrm{Es}_{A,\Lambda}(x), \qquad (3.87)$$

   or, in the matrix form, $u_{A,\Lambda} = G_\Lambda\,\mathrm{Es}_{A,\Lambda}$.
(ii) Let us define

$$\mathrm{cap}_\Lambda(A) = \sum_{x\in\Lambda} \mathrm{Es}_{A,\Lambda}(x). \qquad (3.88)$$

   What general properties of this notion of capacity can one obtain?
(iii) Now, consider the case $\Lambda = B(R)$, and let R grow to infinity. What happens then to the objects we just considered? How does the "canonical" two-dimensional capacity (as defined in (3.18)) then relates to the one defined in (3.88)?

Exercise 3.41. Lemma 3.11 permits us to obtain good approximations for probabilities of exiting annuli at inner/outer boundaries, in the case when the two circumferences are (almost) concentric. But what can be said about the situation depicted on Figure 3.8 (of course, in the regime when the radii of the circumferences are large)? Can you propose a method of obtaining a good approximation for that probability?

Exercise 3.42. Prove the existence of the potential kernel (defined in the same way as in (3.25)) for the one-dimensional SRW. Can you actually *calculate* it?

Exercise 3.43. Obtain an explicit formula (that is, more explicit than (3.45)) for the one-dimensional Green's function (defined as in (3.43)) restricted on an interval Λ.

4

SRW conditioned on not hitting the origin

So far, the content of this book has been quite "traditional". In this chapter, we finally enter (relatively) unexplored lands. First, we review the classical (however, somewhat underrepresented in the literature) notion of Doob's h-transforms. We then study two-dimensional SRW conditioned on never entering the origin, which is the Doob's h-transform of (unconditional) two-dimensional SRW with respect to its potential kernel a. It turns out that the conditioned walk $\widehat{S}$ is quite an interesting object on its own. As we will see in this chapter, some of its (sometimes surprising) properties include

- $\widehat{S}$ is transient; however,

$$\lim_{y \to \infty} \mathbb{P}_{x_0}[\widehat{S} \text{ ever hits } y] = \frac{1}{2}$$

  for any $x_0 \neq 0$;
- *any* infinite set is recurrent for $\widehat{S}$;
- if A is a "nice" large set (e.g., a large disk or square or segment), then the proportion of sites of A which are ever visited by $\widehat{S}$ is a random variable with approximately uniform distribution on $[0, 1]$.

Also, studying the properties of trajectories of the conditioned walk $\widehat{S}$ will be important for us, since later in Chapter 6 we will make a soup of them.

4.1 Doob's h-transforms

Let us start with a one-dimensional example. Let $(S_n, n \geq 0)$ be the simple random walk in dimension 1. It is well known that for any $0 < x < R$

$$\mathbb{P}_x[\tau_R < \tau_0] = \frac{x}{R} \tag{4.1}$$

– this is the solution of Gambler's Ruin Problem (found in most elementary probability books) for players of equal strength. For the purposes of

73

this chapter, however, it is also important to notice that the preceding fact follows, in a quite straightforward way, from the optional stopping theorem applied to the martingale (S_n) and the stopping time $\tau_0 \wedge \tau_R$. Now, how will the walk behave if we *condition* it to reach R before reaching the origin? Using (4.1), we write

$$
\begin{aligned}
\mathbb{P}_x[S_1 &= x + 1 \mid \tau_R < \tau_0] \\
&= \frac{\mathbb{P}_x[S_1 = x + 1, \tau_R < \tau_0]}{\mathbb{P}_x[\tau_R < \tau_0]} \\
&= \frac{\mathbb{P}_x[S_1 = x + 1]\mathbb{P}_x[\tau_R < \tau_0 \mid S_1 = x + 1]}{\mathbb{P}_x[\tau_R < \tau_0]} \\
&= \frac{\frac{1}{2}\mathbb{P}_{x+1}[\tau_R < \tau_0]}{\mathbb{P}_x[\tau_R < \tau_0]} \\
&= \frac{\frac{1}{2} \times \frac{x+1}{R}}{\frac{x}{R}} \\
&= \frac{1}{2} \times \frac{x + 1}{x},
\end{aligned}
$$

which also implies that $\mathbb{P}_x[S_1 = x - 1 \mid \tau_R < \tau_0] = \frac{1}{2} \times \frac{x-1}{x}$. Notice that, by the way, the *drift* at x of the conditioned walk is of order $\frac{1}{x}$ – this is the so-called Lamperti's process which we have already seen in this book: recall Exercises 2.26 and 2.27. The preceding calculation does not yet formally show that the conditioned walk is a Markov process (strictly speaking, we would have needed to condition on the whole history, to begin with), but let us forget about that for now, and examine the *new* transition probabilities we just obtained, $\hat{p}(x, x - 1) = \frac{1}{2} \times \frac{x-1}{x}$ and $\hat{p}(x, x + 1) = \frac{1}{2} \times \frac{x+1}{x}$. First, it is remarkable that they do not depend on R, which *suggests* that we can just send R to infinity and obtain "the random walk conditioned on never returning to the origin". Secondly, just look at the arguments of $\hat{p}$'s and the second fraction in the right-hand sides: these new transition probabilities are related to the old ones (which are $p(x, y) = \frac{1}{2}$ for $x \sim y$) in a special way:

$$
\hat{p}(x, y) = p(x, y) \times \frac{h(y)}{h(x)} \tag{4.2}
$$

with $h(x) = |x|$ (soon it will be clear why do we prefer to keep the function nonnegative). What is special about this function h is that it is harmonic outside the origin, so that $h(S_{n \wedge \tau_0})$ is a martingale. It is *precisely* this fact that permitted us to obtain (4.1) with the help of the optional stopping theorem.

Keeping the preceding discussion in mind, let us spend some time with generalities. Consider a countable Markov chain on a state space Σ, and let $A \subset \Sigma$ be finite. Let $h : \Sigma \to \mathbb{R}_+$ be a nonnegative function which is zero on A and strictly positive and harmonic outside A, i.e., $h(x) = \sum_y p(x, y)h(y)$ for all $x \notin A$. We assume also that $h(x) \to \infty$ as $x \to \infty$; this clearly implies that the Markov chain is recurrent (recall Theorem 2.4).

Definition 4.1. The new Markov chain with transition probabilities defined as in (4.2) is called *Doob's h-transform* of the original Markov chain with respect to h.

Observe that the harmonicity of h implies that $\hat{p}$'s are transition probabilities indeed:

$$\sum_y \hat{p}(x, y) = \frac{1}{h(x)} \sum_y p(x, y)h(y) = \frac{1}{h(x)} \times h(x) = 1.$$

To the best of the author's knowledge, this kind of object first appeared in [37], in the continuous-space-and-time context. In this book, we again (as in Chapter 3) do not try build a comprehensive general theory of Doob's *h*-transforms, but rather only look at the simplified setup (only simple random walks) and take our profits. Further information can be found e.g. in [19, 64, 103], and the book [38] provides a systematic treatment of the subject in full generality.

Note the following simple calculation: for any $x \notin A \cup \partial_e A$, we have (note that $h(y) \neq 0$ for all $y \sim x$ then)

$$\mathbb{E}_x \frac{1}{h(\widehat{X}_1)} = \sum_{y \sim x} \hat{p}(x, y) \frac{1}{h(y)}$$

$$= \sum_{y \sim x} p(x, y) \frac{h(y)}{h(x)} \frac{1}{h(y)}$$

$$= \frac{1}{h(x)} \sum_{y \sim x} p(x, y)$$

$$= \frac{1}{h(x)},$$

which implies the following:

Proposition 4.2. *The process* $1/h(\widehat{X}_{n \wedge \tau_{A \cup \partial_e A}})$ *is a martingale and the Markov chain* $\widehat{X}$ *is transient.*

(The last statement follows from Theorem 2.5 since $h(x) \to \infty$ as $x \to \infty$.)

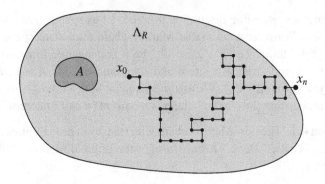

Figure 4.1 Comparing the weights of the path.

Now let us try to get an idea about what the h-transformed chain really does. For technical reasons, let us make another assumption[1]: there exists $c > 0$ such that $|h(x) - h(y)| \leq c$ for all $x \sim y$ (for general Markov chains, $x \sim y$ means $p(x, y) + p(y, x) > 0$).

For $R > 0$, let us define

$$\Lambda_R = \{x \in \Sigma : h(x) \leq R\};$$

under the previous assumptions, Λ_R is finite for any R. Note that the optional stopping theorem implies that, for $x_0 \in \Lambda_R \setminus A$

$$h(x_0) = \mathbb{P}_{x_0}[\tau_{\Lambda_R^c} < \tau_A]\mathbb{E}_{x_0}(h(X_{\tau_{\Lambda_R^c}}) \mid \tau_{\Lambda_R^c} < \tau_A),$$

(recall that $\mathbb{E}_{x_0}(h(X_{\tau_A}) \mid \tau_A < \tau_{\Lambda_R^c}) = 0$ because h is zero on A) and, since the second factor in the preceding display is in $[R, R + c]$, we have

$$\mathbb{P}_{x_0}[\tau_{\Lambda_R^c} < \tau_A] = \frac{h(x_0)}{R}(1 + O(R^{-1})). \qquad (4.3)$$

Then, we consider another countable Markov chain $\widehat{X}$ on the state space $\Sigma \setminus A$ with transition probabilities $\hat{p}(\cdot, \cdot)$ defined as in (4.2) for $x \notin A$. Now, consider a *path* $\wp = (x_0, \ldots, x_{n-1}, x_n)$, where $x_0, \ldots, x_{n-1} \in \Lambda_R \setminus A$ and $x_n \in \Sigma \setminus \Lambda_R$ (here, "path" is simply a sequence of neighbouring sites; in particular, it need not be self-avoiding). The original weight of that path (i.e., the probability that the Markov chain X follows it starting from x_0) is

$$P_\wp = p(x_0, x_1)p(x_1, x_2) \ldots p(x_{n-1}, x_n),$$

---

[1] One can live without this assumption, e.g., in the one-dimensional nearest-neighbour case; see Exercises 4.3 and 4.4.

and the weight of the path for the new Markov chain $\widehat{X}$ will be

$$\widehat{P}_\wp = p(x_0, x_1)\frac{h(x_1)}{h(x_0)}p(x_1, x_2)\frac{h(x_2)}{h(x_1)} \ldots p(x_{n-1}, x_n)\frac{h(x_n)}{h(x_{n-1})}$$

$$= p(x_0, x_1)p(x_1, x_2)\ldots p(x_{n-1}, x_n)\frac{h(x_n)}{h(x_0)}$$

$$= P_\wp \frac{h(x_n)}{h(x_0)}. \tag{4.4}$$

Here comes the key observation: the last term in (4.4) actually equals $\frac{R}{h(x_0)}(1 + O(R^{-1}))$, that is, it is *almost* inverse of the expression in the right-hand side of (4.3). So, we have

$$\widehat{P}_\wp = \frac{P_\wp}{\mathbb{P}_{x_0}[\tau_{\Lambda_R^\complement} < \tau_A]}(1 + O(R^{-1})),$$

that is, the probability that the $\widehat{X}$ chain follows a path is almost the *conditional* probability that that the original chain X follows that path, under the condition that it goes out of Λ_R before reaching A (and the relative error goes to 0 as $R \to \infty$). Now, the (decreasing) sequence of events $\{\tau_{\Lambda_R^\complement} < \tau_A\}$ converges to $\{\tau_A = \infty\}$ as $R \to \infty$. Therefore, we can rightfully call $\widehat{X}$ the Markov chain *conditioned* on never reaching A, even though the probability of the latter event equals zero.

We end this section with an unexpected[2] remark. Recall the conditioned one-dimensional SRW we just constructed: denoting $\Delta_x = S_1 - x$, we calculated that $\mathbb{E}_x\Delta_x = \frac{1}{x}$; and also, obviously, it holds that $\mathbb{E}_x\Delta_x^2 = 1$. So,

$$x\mathbb{E}_x\Delta_x = \mathbb{E}_x\Delta_x^2;$$

notice, by the way, that this relation remains unaffected if one rescales the space by a constant factor. Now recall Exercise 2.26: the preceding equality will (asymptotically) hold in three dimensions (and *only* in three dimensions) for the norm of the SRW (i.e., its distance to the origin). This *suggests* that there may be some "hidden relationship" between the one-dimensional conditioned SRW, and the norm of the three-dimensional SRW. Since in the continuous limit such relationships often reveal themselves better, one may wonder if the (suitably defined) one-dimensional conditioned (on not hitting the origin) Brownian motion is somehow related to the norm of the three-dimensional Brownian motion. The reader is invited to look at this question more closely.

---

[2] Well, maybe quite the contrary.

4.2 Conditioned SRW in two dimensions: basic properties

As you probably expected, we now turn our attention to the two-dimensional SRW. By (3.33), the potential kernel a is ready to play the role of the h, so let us define another random walk $(\widehat{S}_n, n \geq 0)$ on $\mathbb{Z}^2 \setminus \{0\}$ in the following way: its transition probability matrix equals

$$
\hat{p}(x, y) = \begin{cases} \dfrac{a(y)}{4a(x)}, & \text{if } x \sim y, x \neq 0, \\[2mm] 0, & \text{otherwise.} \end{cases} \tag{4.5}
$$

The discussion of the previous section then means that the random walk $\widehat{S}$ is the Doob h-transform of the simple random walk, under the condition of not hitting the origin. Let $\hat{\tau}$ and $\hat{\tau}^+$ be the entrance and the hitting times for $\widehat{S}$; they are defined as in (1.1) and (1.2), only with $\widehat{S}$. We summarize the basic properties of the random walk $\widehat{S}$ in the following:

Proposition 4.3. *The following statements hold:*

(i) *The walk $\widehat{S}$ is reversible, with the reversible measure $\mu(x) = a^2(x)$.*
(ii) *In fact, it can be represented as a random walk on the two-dimensional lattice with the set of conductances $(a(x)a(y), x, y \in \mathbb{Z}^2, x \sim y)$.*
(iii) *The process $1/a(\widehat{S}_{n \wedge \hat{\tau}_N})$ is a martingale.*[3]
(iv) *The walk $\widehat{S}$ is transient.*

Proof Indeed, for (i) and (ii) note that

$$
a^2(x)\hat{p}(x, y) = \frac{a(x)a(y)}{4} = a^2(y)\hat{p}(y, x)
$$

for all adjacent $x, y \in \mathbb{Z}^2 \setminus \{0\}$, and, since a is harmonic outside the origin,

$$
\frac{a(x)a(y)}{\sum_{z \sim x} a(x)a(z)} = \frac{a(y)}{4 \sum_{z \sim x} \frac{1}{4} a(z)} = \frac{a(y)}{4a(x)} = \hat{p}(x, y).
$$

Items (iii) and (iv) are Proposition 4.2.   □

Next, we relate the probabilities of certain events for the walks S and $\widehat{S}$. For $D \subset \mathbb{Z}^2$, let $\Gamma_D^{(x)}$ be the set of all nearest-neighbour finite trajectories that start at $x \in D \setminus \{0\}$ and end when entering ∂D for the first time; denote also $\Gamma_{y,R}^{(x)} = \Gamma_{B(y,R)}^{(x)}$. For $\mathcal{H} \subset \Gamma_D^{(x)}$ write $S \in \mathcal{H}$ (respectively, $\widehat{S} \in \mathcal{H}$) if there exists k such that $(S_0, \ldots, S_k) \in \mathcal{H}$ (respectively, $(\widehat{S}_0, \ldots, \widehat{S}_k) \in \mathcal{H}$). In the next result, we show that $\mathbb{P}_x[S \in \cdot \mid \tau_0 > \tau_{\partial B(R)}]$ and $\mathbb{P}_x[\widehat{S} \in \cdot]$ are almost

---

[3] Recall that $\mathcal{N}$ is the set of the four neighbours of the origin.

indistinguishable on $\Gamma_{0,R}^{(x)}$ (that is, the conditional law of S almost coincides with the unconditional law of $\widehat{S}$).

Lemma 4.4. *Let* $x \in B(R) \setminus \{0\}$, *and assume* $\mathcal{H} \subset \Gamma_{0,R}^{(x)}$. *We have*

$$\mathbb{P}_x[S \in \mathcal{H} \mid \tau_0 > \tau_{\partial B(R)}] = \mathbb{P}_x[\widehat{S} \in \mathcal{H}](1 + O((R \ln R)^{-1})). \tag{4.6}$$

On sets which are distant from the origin, however, we show that S and $\widehat{S}$ have almost the same behaviour (without any conditioning):

Lemma 4.5. *Suppose that* $0 \notin D$, *and assume that* $\mathcal{H} \subset \Gamma_D^{(x)}$; *also, denote* $s = \mathrm{dist}(0, D)$, $r = \mathrm{diam}(D)$. *Then, for* $x \in D$,

$$\mathbb{P}_x[S \in \mathcal{H}] = \mathbb{P}_x[\widehat{S} \in \mathcal{H}]\left(1 + O\left(\frac{r}{s \ln s}\right)\right). \tag{4.7}$$

Proof of Lemmas 4.4 and 4.5 Let us prove (4.6). Assume without loss of generality that no trajectory from $\mathcal{H}$ passes through the origin. Note that for any path $\varrho = (x_0, x_1, \dots, x_n)$ in $\mathbb{Z}^2 \setminus \{0\}$ we have (as in (4.4))

$$\mathbb{P}_{x_0}[\widehat{S}_1 = x_1, \dots, \widehat{S}_n = x_n] = \frac{a(x_1)}{4a(x_0)} \times \frac{a(x_2)}{4a(x_1)} \times \cdots \times \frac{a(x_n)}{4a(x_{n-1})}$$

$$= \frac{a(x_n)}{a(x_0)}\left(\frac{1}{4}\right)^n, \tag{4.8}$$

and therefore it holds that

$$\mathbb{P}_x[\widehat{S} \in \mathcal{H}] = \sum_{\varrho \in \mathcal{H}} \frac{a(\varrho_{\mathrm{end}})}{a(x)}\left(\frac{1}{4}\right)^{|\varrho|},$$

where $|\varrho|$ is the length of ϱ and ϱ_{end} is the last site of ϱ. On the other hand, by Lemma 3.12

$$\mathbb{P}_x[S \in \mathcal{H} \mid \tau_0 > \tau_{\partial B(R)}] = \frac{a(R) + O(R^{-1})}{a(x)} \sum_{\varrho \in \mathcal{H}} \left(\frac{1}{4}\right)^{|\varrho|}.$$

Since $\varrho_{\mathrm{end}} \in \partial B(R)$, we have $a(\varrho_{\mathrm{end}}) = a(R) + O(R^{-1})$ by (3.38), and so (4.6) follows.

The proof of (4.7) is essentially the same, only this time the factor $\frac{a(\varrho_{\mathrm{end}})}{a(x)}$ will be equal to $1 + O(\frac{r}{s \ln s})$, again due to (3.38). □

Next, we need an analogue of Lemma 3.11 for the conditioned random walk, i.e., we would like to estimate the probability that the $\widehat{S}$-walk escapes an annulus through its outer boundary:

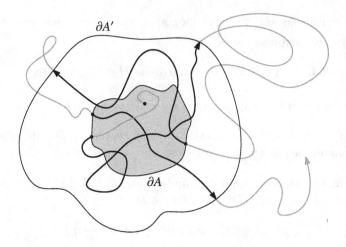

Figure 4.2 Excursions (pictured as bold pieces of trajectories) of random walks between ∂A and $\partial A'$.

Lemma 4.6. *For all $x \in \mathbb{Z}^2$ and $R > r > 0$ such that $x \in B(R) \setminus B(r)$, we have*

$$\mathbb{P}_x[\hat{\tau}_{\partial B(R)} < \hat{\tau}_{B(r)}] = 1 - \frac{(a(x))^{-1} - (a(R) + O(R^{-1}))^{-1}}{(a(r) + O(r^{-1}))^{-1} - (a(R) + O(R^{-1}))^{-1}}, \qquad (4.9)$$

as $r, R \to \infty$.

Proof This is an argument of the type we have seen already many times in this book: use the optional stopping theorem for the martingale $1/(\widehat{S}_{k \wedge \hat{\tau}_N})$ with the stopping time $\hat{\tau}_{\partial B(R)} \wedge \hat{\tau}_{B(r)}$. We let the reader fill the details. □

Letting $R \to \infty$ in (4.9) yields

Corollary 4.7. *Assume $r \geq 1$ and $\|x\| \geq r + 1$. We have*

$$\mathbb{P}_x[\hat{\tau}_{B(r)} = \infty] = 1 - \frac{a(r) + O(r^{-1})}{a(x)}. \qquad (4.10)$$

One typically needs relations such as (4.9) and (4.10) when dealing with *excursions* of the conditioned random walk. If $A \subset A'$ are (finite) sub-sets of $\mathbb{Z}^2$, then the *excursions* between ∂A and $\partial A'$ are pieces of nearest-neighbour trajectories that begin on ∂A and end on $\partial A'$; see Figure 4.2, which is, hopefully, self-explanatory. We refer to section 3.4 of [26] for formal definitions.

For example, let $A = B(r)$ and $A' = B(2r)$. Then, for each $x \in \partial A'$, (4.10) and (3.36) imply that

$$P_x[\hat{\tau}_{B(r)} = \infty] = 1 - \frac{\frac{2}{\pi}\ln r + \gamma' + O(r^{-1})}{\frac{2}{\pi}\ln 2r + \gamma' + O(r^{-1})}$$

$$= \frac{\ln 2 + O(r^{-1})}{\ln r + (\frac{\pi}{2}\gamma' + \ln 2) + O(r^{-1})},$$

that is, the distribution of the total number of such excursions is *very close* to geometric with success probability $\frac{\ln 2}{\ln r + (\frac{\pi}{2}\gamma' + \ln 2)}$. Arguments of this sort will appear several times in the subsequent exposition.

What we do next is develop *some* potential theory for the conditioned walk $\widehat{S}$.

4.3 Green's function and capacity

Green's function of the conditioned walk is defined in the following way (completely analogous to (3.2)): for $x, y \in \mathbb{Z}^2 \setminus \{0\}$

$$\widehat{G}(x, y) = \mathbb{E}_x \sum_{k=0}^{\infty} \mathbb{1}\{\widehat{S}_k = y\}. \tag{4.11}$$

It is remarkable that one is actually able to *calculate* this function in terms of the potential kernel a (this is theorem 1.1 of [77]):

Theorem 4.8. *For all $x, y \in \mathbb{Z}^2 \setminus \{0\}$ it holds that*

$$\widehat{G}(x, y) = \frac{a(y)}{a(x)}(a(x) + a(y) - a(x - y)). \tag{4.12}$$

Proof First, we need a very simple general fact about hitting times of recurrent Markov chains:

Lemma 4.9. *Let (X_n) be a recurrent Markov chain on a state space Σ, and $x \in \Sigma$, $A, B \subset \Sigma$ are such that $A \cap B = \emptyset$ and $x \notin A \cup B$. Then*

$$P_x[\tau_A < \tau_B] = P_x[\tau_A < \tau_B \mid \tau_x^+ > \tau_{A \cup B}] \tag{4.13}$$

(that is, the events $\{\tau_A < \tau_B\}$ and $\{\tau_x^+ > \tau_{A \cup B}\}$ are independent under P_x).

Proof Informally (see Figure 4.3): let $p := P_x[\tau_A < \tau_B \mid \tau_x^+ > \tau_{A \cup B}]$ be the value of the probability in the right-hand side of (4.13). At the moments when the walker visits x, it tosses a coin to decide if it will revisit it before coming to $A \cup B$, or not. When it decides to definitely leave x for $A \cup B$, the probability of choosing A is p, so it is p overall. Making this argument rigorous and boring is left as an exercise. $\square$

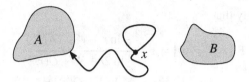

Figure 4.3 On the proof of Lemma 4.9.

We continue proving Theorem 4.8. Fix (a large) $R > 0$, abbreviate $\Lambda_R = B(R) \setminus \{0\}$, and let us denote for $y \in \Lambda_R$

$$N^*_{y,R} = \sum_{k=0}^{\tau_{\Lambda_R^c}} \mathbf{1}\{S_k = y\},$$

$$\widehat{N}^*_{y,R} = \sum_{k=0}^{\hat{\tau}_{\Lambda_R^c}} \mathbf{1}\{\widehat{S}_k = y\},$$

to be the numbers of visits to y before hitting 0 or $\partial_e B(R)$, for the SRW and the conditioned SRW. Let us also denote $\widehat{G}_R(x,y) = \mathbb{E}_x \widehat{N}^*_{y,R}$. Before looking at the next argument, it is a good idea to recall how "exactly k visits" became "at least k visits" in (3.12). Now, let $x \in \Lambda_R$ and observe that, on one hand,

$$\mathbb{P}_x[N^*_{y,R} = n, \tau_{\partial_e B(R)} < \tau_0]$$
$$= \mathbb{P}_x[N^*_{y,R} \geq n]\mathbb{P}_y[\tau_{\partial_e B(R)} < \tau_0, \tau_y^+ > \tau_{\Lambda_R^c}]$$
$$= \mathbb{P}_x[N^*_{y,R} \geq n]\mathbb{P}_y[\tau_y^+ > \tau_{\Lambda_R^c}]\mathbb{P}_y[\tau_{\partial_e B(R)} < \tau_0 \mid \tau_y^+ > \tau_{\Lambda_R^c}]$$

(by Lemma 4.9)

$$= \mathbb{P}_x[N^*_{y,R} \geq n]\mathbb{P}_y[\tau_y^+ > \tau_{\Lambda_R^c}]\mathbb{P}_y[\tau_{\partial_e B(R)} < \tau_0]$$
$$= \mathbb{P}_x[N^*_{y,R} = n]\mathbb{P}_y[\tau_{\partial_e B(R)} < \tau_0]$$

(by Lemma 3.12)

$$= \mathbb{P}_x[N^*_{y,R} = n]\frac{a(y)}{a(R) + O(R^{-1})}, \tag{4.14}$$

and, on the other hand, the same expression can be also treated in the following way:

$$\mathbb{P}_x[N^*_{y,R} = n, \tau_{\partial_e B(R)} < \tau_0]$$
$$= \mathbb{P}_x[N^*_{y,R} = n \mid \tau_{\partial_e B(R)} < \tau_0]\mathbb{P}_x[\tau_{\partial_e B(R)} < \tau_0]$$

(by Lemma 4.4)

$$= \mathbb{P}_x[\widehat{N}^*_{y,R} = n](1 + O((R \ln R)^{-1}))\mathbb{P}_x[\tau_{\partial_e \mathrm{B}(R)} < \tau_0]$$

(again, by Lemma 3.12)

$$= \mathbb{P}_x[\widehat{N}^*_{y,R} = n](1 + O((R \ln R)^{-1}))\frac{a(x)}{a(R) + O(R^{-1})}. \qquad (4.15)$$

Note also that $a(R) + O(R^{-1}) = a(R)(1 + O((R \ln R)^{-1}))$. So, since (4.14) and (4.15) must be equal, we have

$$a(x)\mathbb{P}_x[\widehat{N}^*_{y,R} = n] = a(y)\mathbb{P}_x[N^*_{y,R} = n](1 + O((R \ln R)^{-1}));$$

multiplying by n and summing in $n \geq 1$, we obtain

$$a(x)\widehat{G}_R(x, y) = a(y)G_{\Lambda_R}(x, y)(1 + O((R \ln R)^{-1})). \qquad (4.16)$$

Note that $\widehat{G}_R(x, y) \to \widehat{G}(x, y)$ as $R \to \infty$, due to the Monotone Convergence Theorem. Next, we are actually able to say something about $G_{\Lambda_R}(x, y)$: by Theorem 3.13, it holds that[4]

$$G_{\Lambda_R}(x, y) = \mathbb{E}_x a(S_{\tau_{\Lambda_R^C}} - y) - a(x - y)$$

(once again, by Lemma 3.12)

$$= \frac{a(x)}{a(R) + O(R^{-1})}(a(R) + O(\tfrac{\|y\|+1}{R}))$$
$$+ \left(1 - \frac{a(x)}{a(R) + O(R^{-1})}\right)a(y) - a(x - y)$$
$$= a(x) + a(y) - a(x - y) + O(\tfrac{\|y\|+1}{R} + \tfrac{a(x)a(y)}{a(R)}).$$

Inserting this back to (4.16) and sending R to infinity, we obtain (4.12). $\quad\square$

At this point, let us recall that the function $1/a(\cdot)$ is harmonic[5] on $\mathbb{Z}^2 \setminus (\mathcal{N} \cup \{0\})$, and observe that Green's function $\widehat{G}(\cdot, y)$ is harmonic on $\mathbb{Z}^2 \setminus \{0, y\}$ (as before, this is an immediate consequence of the total expectation formula; recall (3.3)). It turns out that this "small" difference will be quite important: indeed, the latter fact will be operational in some places later in this chapter, for applying the optional stopping theorem in some particular settings. For future reference, we formulate the preceding fact in the equivalent form:

---

[4] By the way, recall the solution of Exercise 3.34.
[5] With respect to the conditioned walk.

Proposition 4.10. *For any $y \in \mathbb{Z}^2 \setminus \{0\}$, the process $(\widehat{G}(S_{n \wedge \hat{\tau}_y}, y), n \geq 0)$ is a martingale. Moreover, let us define*

$$\hat{\ell}(x, y) = 1 + \frac{a(y) - a(x - y)}{a(x)} = \frac{\widehat{G}(x, y)}{a(y)}. \tag{4.17}$$

Then the process $(\hat{\ell}(\widehat{S}_{n \wedge \hat{\tau}_y}, y), n \geq 0)$ is a martingale.

By the way, notice that

$$\lim_{x \to \infty} \hat{\ell}(x, y) = 0 \tag{4.18}$$

for any fixed y, so the last process is a "martingale vanishing at infinity", which makes it more convenient for applications via the optional stopping theorem (so this is why we kept "1+" in (4.17)).

Now, as we already know from Section 3.1, it is possible to obtain exact expressions (in terms of Green's function) for one-site escape probabilities, and probabilities of (not) hitting a given site. Indeed, since, under $\mathbb{P}_x$, the number of visits (counting the one at time 0) to x is geometric with success probability $\mathbb{P}_x[\hat{\tau}_x = \infty]$, using (4.13), we obtain

$$\mathbb{P}_x[\hat{\tau}_x^+ < \infty] = 1 - \frac{1}{\widehat{G}(x, x)} = 1 - \frac{1}{2a(x)} \tag{4.19}$$

for $x \neq 0$. Also, quite analogously to (3.4), since

$$\widehat{G}(x, y) = \mathbb{P}_x[\hat{\tau}_y^+ < \infty]\widehat{G}(y, y) \qquad \text{for } x \neq y, \, x, y \neq 0$$

(one needs to go to y first, to start counting visits there), we have

$$\mathbb{P}_x[\hat{\tau}_y < \infty] = \frac{\widehat{G}(x, y)}{\widehat{G}(y, y)} = \frac{a(x) + a(y) - a(x - y)}{2a(x)}. \tag{4.20}$$

Let us also observe that (4.20) implies the following surprising fact: for any $x \neq 0$,

$$\lim_{y \to \infty} \mathbb{P}_x[\hat{\tau}_y < \infty] = \frac{1}{2}. \tag{4.21}$$

It is interesting to note that this fact permits us to obtain a criterion for recurrence of a set with respect to the conditioned walk. Quite analogously to Definition 3.5, we say that a set is recurrent with respect to a (transient) Markov chain, if it is visited infinitely many times almost surely; a set is called *transient*, if it is visited only finitely many times almost surely. Recall that, for SRW in dimension $d \geq 3$, the characterization is provided by *Wiener's criterion* (Theorem 3.7) formulated in terms of capacities of intersections of the set with exponentially growing annuli. Although this result does provide a complete classification, it may be difficult to apply it in practice, because it is not always trivial to calculate (even to *estimate*) capacities. Now, it turns out that for the conditioned two-dimensional walk $\widehat{S}$ the characterization of recurrent and transient sets is particularly simple:

Theorem 4.11. *A set $A \subset \mathbb{Z}^2$ is recurrent with respect to $\widehat{S}$ if and only if A is infinite.*

Proof of Theorem 4.11   Clearly, we only need to prove that every infinite subset of $\mathbb{Z}^d$ is recurrent for $\widehat{S}$. As mentioned before, this is basically a consequence of (4.21). Indeed, let $\widehat{S}_0 = x_0$; since A is infinite, by (4.21) one can find $y_0 \in A$ and R_0 such that $\{x_0, y_0\} \subset \mathrm{B}(R_0)$ and

$$\mathbb{P}_{x_0}[\hat{\tau}_{y_0} < \hat{\tau}_{\partial \mathrm{B}(R_0)}] \geq \frac{1}{3}.$$

Then, for any $x_1 \in \partial \mathrm{B}(R_0)$, we can find $y_1 \in A$ and $R_1 > R_0$ such that $y_1 \in \mathrm{B}(R_1) \setminus \mathrm{B}(R_0)$ and

$$\mathbb{P}_{x_1}[\hat{\tau}_{y_1} < \hat{\tau}_{\partial \mathrm{B}(R_1)}] \geq \frac{1}{3}.$$

Continuing in this way, we can construct a sequence $R_0 < R_1 < R_2 < \ldots$ (depending on the set A) such that, for each $k \geq 0$, the walk $\widehat{S}$ hits A on its way from $\partial \mathrm{B}(R_k)$ to $\partial \mathrm{B}(R_{k+1})$ with probability at least $\frac{1}{3}$, regardless of the past. This clearly implies that A is a recurrent set.   □

Next, following the script of Section 3.1, let us discuss the notion of capacity for the conditioned walk. It is tempting to just repeat the previous definitions by reusing (3.10), but, at this point, some care has to be taken. Namely, recall that in Section 3.1 we used path reversals a few times – for SRW, the probability that it follows a path is the same as the probability that it follows the reversed path. This is no longer true for the conditioned walk; it is still reversible, but the reversible measure is not constant (recall Proposition 4.3). As a consequence, we have that Green's function (4.12) for the conditioned walk is no longer symmetric; instead, it holds that

$$a^2(x)\widehat{G}(x, y) = a^2(y)\widehat{G}(y, x) \tag{4.22}$$

for all $x, y \in \mathbb{Z}^2 \setminus \{0\}$. This, of course, follows directly from (4.12), but can be also obtained independently, analogously to the proof of (3.44) (recall also Exercise 3.27).

For finite $A \subset \mathbb{Z}^d$ and $x \in \mathbb{Z}^d$, quite analogously to (3.9), let us denote by

$$\widehat{\mathrm{Es}}_A(x) = \mathbb{P}_x[\hat{\tau}_A^+ = \infty]\mathbf{1}\{x \in A\} \tag{4.23}$$

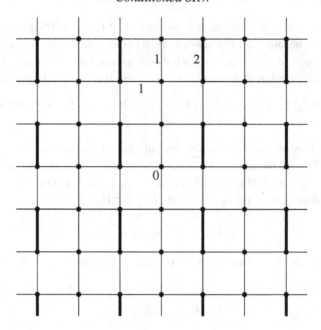

Figure 4.4 Conductances of the "thicker" edges are equal to 2, and the conductances of all other edges (including the ones orthogonal to this book, which are not shown on the picture) are 1.

the escape probability from $x \in A$ with respect to the conditioned walk. The crucial observation (that the reader is strongly invited to check) is that one can show that, for any $A \subset \mathbb{Z}^2 \setminus \{0\}$,

$$\mathbb{P}_x[\hat{\tau}_A < \infty] = \sum_{y \in A} \widehat{G}(x, y)\,\widehat{\mathrm{Es}}_A(y) = \sum_{y \in \mathbb{Z}^d} \widehat{G}(x, y)\,\widehat{\mathrm{Es}}_A(y) \qquad (4.24)$$

exactly in the same way as (3.11) was proved!

However, maybe somewhat surprisingly, it is *not* a good idea to define the capacity for the conditioned walk as in (3.10). To explain this, consider first a *toy model*. Let X be the random walk on the three-dimensional integer lattice with conductances on all the horizontal planes defined as in Figure 4.4, and the conductances of all vertical edges being equal to 1. Clearly, X is transient and reversible, with the reversible measure

$$\mu(x) = \begin{cases} 6, & \text{if } x \cdot e_1 \text{ is even,} \\ 7, & \text{if } x \cdot e_1 \text{ is odd.} \end{cases}$$

This informally means that the odds that the process is in a site with an odd abscissa is $\frac{6}{6+7} = \frac{6}{13}$. Now, consider the n-step transition probability $p^{(n)}(x, y)$, where n is of the same parity as $x - y$. Intuitively, it should not depend so much on x (since the walk normally "forgets" its initial point anyway), but there should be a *substantial* dependence on y: by the preceding discussion, if, for example, $y \cdot e_1$ is even, the ratio $\frac{p^{(n)}(x,y)}{p^{(n)}(x,y')}$ should be close to $\frac{6}{7}$ in case when $y' \cdot e_1$ is odd and y' has the same parity as y and is "not far away" from it. So, if we divide $p^{(n)}(x, y)$ by $\mu(y)$, this will (almost) remove the dependence on the second argument; since Green's function is the sum of $p^{(n)}$'s, $G(x, y)/\mu(y)$ looks like the "right" object to consider (it "should" depend on the distance between x and y, but not so much on x and y themselves). Then an analogue of (3.11) would be

$$\mathbb{P}_x[\tau_A < \infty] = \sum_{y \in A} G(x, y) \, \mathrm{Es}_A(y) = \sum_{y \in A} \frac{G(x, y)}{\mu(y)} \times \mu(y) \, \mathrm{Es}_A(y),$$

so, at least if A is finite and the starting point x is far away from A, the probability of eventually hitting A would be a product of a factor which (almost) only depends on the distance with $\sum_{y \in A} \mu(y) \, \mathrm{Es}_A(y)$. This *indicates* that the last quantity might be the *correct* definition of the capacity.

It is important to notice, however, that the preceding *correct* definition is, in principle, *non-unique*: as we remember from Section 2.2, the reversible measure is not unique (one can always multiply it by a positive constant). The same is also true with respect to the conductances: if we multiply all of them by the same positive constant, the corresponding random walk will remain the same. Still, we will see later that there is a *canonical* way to define the capacity for conditioned random walks (i.e., we can choose that multiplicative constant in a *natural* way).

Let us go back to the conditioned walk $\widehat{S}$. First, note that, by (3.38),

$$\widehat{G}(x, y) = \frac{a(y)}{a(x)}(a(x) + a(y) - a(x - y)) = \frac{a(y)}{a(x)}(a(y) + O(\tfrac{\|y\|}{\|x\|})) \quad (4.25)$$

as $x \to \infty$ and $\frac{\|y\|}{\|x\|} \to 0$. By (4.12), we have

$$\frac{\widehat{G}(x, y)}{a^2(y)} = \frac{\widehat{G}(y, x)}{a^2(x)} = \frac{a(x) + a(y) - a(x - y)}{a(x)a(y)},$$

and so it is natural to introduce new notation $\hat{g}(x, y) = \frac{\widehat{G}(x,y)}{a^2(y)} = \hat{g}(y, x)$ for the "symmetrized" Green's function. Then (4.25) implies that

$$\hat{g}(x, y) = \frac{\widehat{G}(x, y)}{a^2(y)} = \frac{1}{a(x)}(1 + O(\tfrac{\|y\|}{\|x\| \ln(\|y\|+1)})) \quad (4.26)$$

as $x \to \infty$, which indeed essentially depends on $\|x\|$. Therefore, if we rewrite (4.24) in the following way,

$$\mathbb{P}_x[\hat{\tau}_A < \infty] = \sum_{y \in A} \hat{g}(x, y) \times a^2(y) \, \widehat{\mathrm{Es}}_A(y), \qquad (4.27)$$

we see that the first terms in the previous summation are "almost" the same for large x. According to the preceding discussion, it is then reasonable to adopt the following definition for the capacity $\widehat{\mathrm{cap}}(\cdot)$ with respect to the conditioned walk:

$$\widehat{\mathrm{cap}}(A) = \sum_{y \in A} a^2(y) \, \widehat{\mathrm{Es}}_A(y). \qquad (4.28)$$

Let us go back to the recent observation that, *in principle*, the capacity is defined up to a multiplicative factor. Why is a better than, say, $3a$ for the role of the function h of Section 4.1? How should we choose one of them *canonically*? A reasonable way to do it is the following: h should be such that $\mathbb{E}_0 h(\widehat{S}_1) = 1$; as we know (recall (3.35)), $h \equiv a$ is then the right choice indeed.

Now, we have two notions of two-dimensional capacity: one for the original recurrent walk, and another one for the transient conditioned walk. What is the relationship between them? Remarkably, it is very simple:

Theorem 4.12. *For all $A \subset \mathbb{Z}^2 \setminus \{0\}$, we have*

$$\widehat{\mathrm{cap}}(A) = \mathrm{cap}\,(A \cup \{0\}). \qquad (4.29)$$

Proof Indeed, let us write for $x \in A$

$$\widehat{\mathrm{Es}}_A(x) = \mathbb{P}_x[\hat{\tau}_A^+ = \infty]$$
$$= \lim_{R \to \infty} \mathbb{P}_x[\hat{\tau}_A^+ > \hat{\tau}_{\partial B(R)}]$$

(by Lemma 4.4)

$$= \lim_{R \to \infty} \mathbb{P}_x[\tau_A^+ > \tau_{\partial B(R)} \mid \tau_{\partial B(R)} < \tau_0]$$

(by Lemma 3.12)

$$= \lim_{R \to \infty} \mathbb{P}_x[\tau_{A \cup \{0\}}^+ > \tau_{\partial B(R)}] \frac{a(R) + O(R^{-1})}{a(x)}$$

(by (3.54))

$$= \frac{\mathrm{hm}_A(x)}{a(x)}. \qquad (4.30)$$

To conclude the proof, just insert (4.30) into (4.28) and then compare the result to the "usual" definition (3.62) of the two-dimensional capacity (with $y_0 = 0$). □

Let us also obtain an expression for the probability of avoiding *any* finite set. The following estimate, however, only works well when the starting point is much farther from the origin than the "target" set; nevertheless, once again it validates the intuition that "capacity measures how big is the set from the walker's point of view". Assume that $A \subset B(r)$, and $\|x\| \geq r+1$. Then, with (4.28) in mind, just look at (4.26) and (4.27) to obtain that

$$\mathbb{P}_x[\hat{\tau}_A < \infty] = \frac{\widehat{\mathrm{cap}}(A)}{a(x)}\left(1 + O\left(\frac{r}{\|x\|\ln(r+1)}\right)\right). \tag{4.31}$$

The remainder of this section will be a bit too technical for the author's taste, but, unfortunately, the lack of translational invariance of the conditioned walk claims its price. The reader certainly remembers that the estimate (3.38) on the two-point differences of the potential kernel was quite instrumental in several arguments of this book; likewise, it will be important to have difference estimates for the function $\hat{g}$ as well:

Lemma 4.13. *Assume that $x, y, z \in \mathbb{Z}^2 \setminus \{0\}$ are distinct and such that* $\|x - y\| \wedge \|x - z\| \geq 5\|y - z\|$. *Then*

$$\left|\hat{g}(x, y) - \hat{g}(x, z)\right| \leq O\left(\frac{\|y-z\|}{\|x-y\|\ln(1+\|x\|\vee\|y\|\vee\|z\|)\ln(1+\|y\|\vee\|z\|)}\right). \tag{4.32}$$

Proof First, let us write

$$\hat{g}(x, y) - \hat{g}(x, z)$$
$$= \frac{a(x) + a(y) - a(x - y)}{a(x)a(y)} - \frac{a(x) + a(z) - a(x - z)}{a(x)a(z)}$$
$$= \frac{a(x)a(z) - a(x - y)a(z) - a(x)a(y) + a(x - z)a(y)}{a(x)a(y)a(z)}$$

(put $\pm a(x - z)a(z)$ to the numerator, then group accordingly)

$$= \frac{a(x)(a(z) - a(y)) - a(x - z)(a(z) - a(y)) + a(z)(a(x - z) - a(x - y))}{a(x)a(y)a(z)}. \tag{4.33}$$

Throughout this proof, let us assume without loss of generality that $\|y\| \geq \|z\|$. Since the walk $\widehat{S}$ is not spatially homogeneous (and, therefore, $\hat{g}$ is not translationally invariant), we need to take into account the relative positions of the three sites with respect to the origin. Specifically, we will consider the following three different cases (see Figure 4.5).

Figure 4.5 On the proof of Lemma 4.13, the three cases to
consider (from left to right): (1) $\|x\|, \|y\|$ are of the same
logarithmic order and $\|x\|$ is not much larger than $\|y\|$, (2) $\|x\|$ is
much smaller than $\|y\|$, and (3) $\|x\|$ is *significantly* larger than $\|y\|$.

Case 1: $\|y\|^{1/2} \le \|x\| \le 2\|y\|$.

In this case, the first thing to note is that

$$\|y - z\| \le \frac{\|x - y\|}{5} \le \frac{\|x\| + \|y\|}{5} \le \frac{2\|y\| + \|y\|}{5} = \frac{3}{5}\|y\|,$$

so $\|z\| \ge \frac{2}{5}\|y\|$, meaning that $\|y\|$ and $\|z\|$ must be of the same order; this
then implies that $a(x), a(y), a(z)$ are all of the same order too. Then we
use (3.38) on the three parentheses in the numerator of (4.33), to obtain after some elementary calculations that the expression there is at most of or-
der $\frac{\|y-z\|}{\|x-y\|} \ln \|y\|$, while the denominator is of order $\ln^3 \|y\|$. This proves (4.32)
in case 1.

Case 2: $\|x\| < \|y\|^{1/2}$.

Here, it is again easy to see that $\|y\|$ and $\|z\|$ must be of the same order.
Now we note that, by (3.38), $a(x - z) = a(z) + O(\frac{\|x\|}{\|y\|})$, so, inserting this
to (4.33) (and also using that $a(y) - a(z) = O(\frac{\|y-z\|}{\|y\|})$), we find that it is equal
to

$$\frac{a(x)(a(z) - a(y)) + a(z)(a(y) - a(z) + a(x - z) - a(x - y)) + O(\frac{\|x\|}{\|y\|} \cdot \frac{\|y-z\|}{\|y\|})}{a(x)a(y)a(z)}$$

$$= \frac{a(z) - a(y)}{a(y)a(z)} + \frac{a(y) - a(z) + a(x - z) - a(x - y)}{a(x)a(y)} + O(\frac{\|x\| \cdot \|y-z\|}{\|y\|^2 \ln^2 \|y\|}). \quad (4.34)$$

Now, by (3.38), the first term is $O(\frac{\|y-z\|}{\|y\| \ln^2 \|y\|})$ (that is, exactly what we need,
since $\|y\|$ and $\|y - x\|$ are of the same order), and the third term is clearly
of smaller order. As for the second term, note that, by (3.36) and using the
fact that $\left|\|x-y\| \cdot \|z\| - \|y\| \cdot \|x-z\|\right| \le O(\|x\| \cdot \|y-z\|)$ by Ptolemy's inequality,

we obtain

$$a(y) - a(z) + a(x - z) - a(x - y)$$
$$= \frac{2}{\pi} \ln \frac{\|y\| \cdot \|x - z\|}{\|x - y\| \cdot \|z\|} + O(\|y\|^{-2})$$
$$= \frac{2}{\pi} \ln \Big(1 - \frac{\|x - y\| \cdot \|z\| - \|y\| \cdot \|x - z\|}{\|x - y\| \cdot \|z\|}\Big) + O(\|y\|^{-2})$$
$$= O\Big(\frac{\|x\| \cdot \|y - z\|}{\|x - y\| \cdot \|z\|} + \|y\|^{-2}\Big) = O\Big(\frac{\|x\| \cdot \|y - z\|}{\|z\|^2}\Big),$$

so it is again of smaller order than the first term. This shows (4.32) in case 2.

Case 3: $\|x\| > 2\|y\|$.

Notice that, in this case, $\|z\|$ need not be of the same order as $\|y\|$; it may happen to be significantly smaller. Here (by also grouping the first two terms in the numerator) we rewrite (4.33) as

$$\frac{(a(x) - a(x - z))(a(z) - a(y))}{a(x)a(y)a(z)} + \frac{a(x - z) - a(x - y)}{a(x)a(y)}. \qquad (4.35)$$

By (3.38), the second term is $O\Big(\frac{\|y - z\|}{\|x - y\| \ln(1 + \|x\|) \ln(1 + \|y\|)}\Big)$ (that is, exactly what we need). Next, observe that (recall that we assumed that $\|y\| \geq \|z\|$)

$$\ln \|y\| - \ln \|z\| = \ln \frac{\|y\|}{\|z\|} \leq \frac{\|y\|}{\|z\|} - 1 = \frac{\|y\| - \|z\|}{\|z\|} \leq \frac{\|y - z\|}{\|z\|}.$$

Therefore (also using (3.38) on the first factor), the numerator of the first term is $O\Big(\frac{\|z\|}{\|x\|} \times \frac{\|y - z\|}{\|z\|}\Big) = O\Big(\frac{\|y - z\|}{\|x\|}\Big)$, and so (since the denominator is not less than $a(x)a(y)$) the first term in (4.35) is at most of the same order as the second one. This concludes the proof of Lemma 4.13. $\qquad \square$

Lemma 4.13 permits us to obtain the following useful expression for the probability of ever hitting A from a distant site (which can be seen as a generalization of (4.31)). Consider a finite $A \subset \mathbb{Z}^2 \setminus \{0\}$ with $y_0 \in A$ and note first that $r_0 := \|y_0\| + \text{diam}(A)$ will be of the same order regardless of the choice of y_0. Assume now that $\text{dist}(x, A) > 5 \, \text{diam}(A)$; then, it also holds that $\|x\| \vee \|y\| \vee \|z\|$ is of the same order as $\|x\| \vee (\|y_0\| + \text{diam}(A))$ for any choice of $y, z \in A$. Indeed, trivially $\|x\| \vee \|y\| \vee \|z\| \leq \|x\| \vee (\|y_0\| + \text{diam}(A))$; on the other hand, $\|y\| + \text{diam}(A) < \|y\| + \text{dist}(x, A) \leq \|y\| + \|x - y\| \leq \|x\| + 2\|y\| \leq 3(\|x\| \vee \|y\| \vee \|z\|)$. Then note that $1 + \|y\| + \|y - z\| \leq 1 + 2\|y\| + \|z\| \leq 3(1 + \|y\| \vee \|z\|) < (1 + \|y\| \vee \|z\|)^3$ (since the expression in parentheses is at least 2), so

$$\frac{\|y - z\|}{\ln(1 + \|y\| \vee \|z\|)} \leq 3 \frac{\|y - z\|}{\ln(1 + \|y\| + \|y - z\|)} \leq 3 \frac{\text{diam}(A)}{\ln(1 + \|y\| + \text{diam}(A))},$$

where the second inequality is due to the fact that the function $f(x) = \frac{x}{\ln(a+x)}$ is increasing on $(0, +\infty)$ for any $a \geq 1$. So, recalling (4.28) and (4.24), we see that Lemma 4.13 gives us (in the case $\mathrm{dist}(x, A) > 5 \, \mathrm{diam}(A)$) that

$$\mathbb{P}_x[\hat{\tau}_A < \infty] = \widehat{\mathrm{cap}}(A)(\hat{g}(x, y_0) + O(\tfrac{\mathrm{diam}(A)}{\mathrm{dist}(x,A)\ln(1+\|x\|\vee r_0)\ln(1+r_0)})). \quad (4.36)$$

So, following Section 3.1, we have now delivered Green's function and the capacity to the conditioned walk; we pass now to the third important ingredient.

4.4 Harmonic measure

As before, we would like to define the harmonic measure as "the entrance measure from infinity". To figure out how it should be defined, recall the informal explanation of the harmonic measure's definition (for the transient case $d \geq 3$) on page 43, just after Theorem 3.8. There is a sentence there which reads, "Since the time reversal does not change the "weight" of the trajectory..."; now, this is no longer true! Since (recall Proposition 4.3 (i)) the conditioned walk is reversible with the reversible measure $(a^2(x), x \in \mathbb{Z}^2 \setminus \{0\})$, the "weight" of the reversed trajectory which starts at $y \in \partial A$ should now be multiplied by $a^2(y)$. So, that heuristic argument suggests that, very similarly to what we have seen in Section 3.1, the harmonic measure should be proportional to y's contribution in the summation (4.28).

Indeed, it turns out to be the case. Let us define the harmonic measure with respect to the conditioned walk in the following way:

$$\widehat{\mathrm{hm}}_A(y) = \frac{a^2(y)\,\widehat{\mathrm{Es}}_A(y)}{\widehat{\mathrm{cap}}(A)} = \frac{a(y)\,\mathrm{hm}_A(y)}{\widehat{\mathrm{cap}}(A)} \quad (4.37)$$

(the last equality holds by (4.30)). We can interpret (4.37) in the following way: $\widehat{\mathrm{hm}}$ is hm biased by a.

Now, the goal will be to prove that, analogously to the harmonic measures considered in Chapter 3, $\widehat{\mathrm{hm}}_A$ is indeed the "entrance measure from infinity". We will aim directly to prove a *quantitative* result, i.e., one that contains explicit estimates on the Radon–Nikodym derivative of the entrance measure from a distant starting point with respect to the harmonic measure. We will see that these estimates match those that we obtained for SRW. However, before we will be able to obtain that result, we will need a couple of technical facts.[6] These are, honestly, a bit tedious to prove, and,

---

[6] The results of the remaining part of this section are taken from [77].

again, they are so because of the lack of the translational invariance of the conditioned walk.

The first technical fact that we need is that the conditioned walk can go out of an annulus with uniformly positive probability:

Lemma 4.14. *Let* b, C *be positive constants such that* $1 + b < C$, *and assume that* $r \geq 1, x_0, y_0 \in \mathbb{Z}^2 \setminus \{0\}$ *are such that* $x_0 \in B(y_0, Cr)$ *and* $\|x_0 - y_0\| > (1 + b)r$. *Then, there exists a constant* $c' > 0$ *(depending* only *on* b *and* C*) such that*

$$\mathbb{P}_{x_0}[\hat{\tau}_{\partial B(y_0, Cr)} < \hat{\tau}_{B(y_0, r)}] \geq c'. \tag{4.38}$$

Proof Note that we can assume that r is large enough, otherwise the uniform ellipticity of the conditioned walk[7] will imply the claim. First of all, it is clear (and can be proved in several possible ways; see Exercise 4.12) that the analogue of (4.38) holds for SRW, i.e., for all r, x_0, y_0 as before, it holds that

$$\mathbb{P}_{x_0}[\tau_{\partial B(y_0, 2Cr)} < \tau_{B(y_0, r)}] \geq c'' \tag{4.39}$$

for some $c'' > 0$ which depends only on b and C. Now, the idea is to derive (4.38) from (4.39). Recall that (as in (4.8)) the weight $\widehat{P}_\varrho$ with respect to $\widehat{S}$ of a finite path ϱ which starts at x_0 and does not pass through the origin equals $\frac{a(\varrho_{end})}{a(x_0)} P_\varrho$, where P_ϱ is the weight of the same path with respect to the SRW. We can then write for any $R > 0$ such that $x_0 \in B(y_0, R)$

$$\mathbb{P}_{x_0}[\hat{\tau}_{\partial B(y_0, R)} < \hat{\tau}_{B(y_0, r)}]$$

$$\geq \min_{z \in \partial B(y_0, R)} \frac{a(z)}{a(x_0)} \times \mathbb{P}_{x_0}[\tau_{\partial B(y_0, R)} < \tau_{B(y_0, r)}, \tau_{\partial B(y_0, R)} < \tau_0]$$

$$\geq \min_{z \in \partial B(y_0, R)} \frac{a(z)}{a(x_0)} \times \left(\mathbb{P}_{x_0}[\tau_{\partial B(y_0, R)} < \tau_{B(y_0, r)}] - \mathbb{P}_{x_0}[\tau_0 < \tau_{\partial B(y_0, R)}]\right). \tag{4.40}$$

Now, a key observation is that, for all large enough r, the property

$$\frac{a(z)}{a(x_0)} \geq \frac{1}{2} \text{ for all } z \in \partial B(y_0, R) \tag{4.41}$$

holds for either $R = Cr$ or $R = 2Cr$ (or both). Indeed, roughly speaking, for (4.41) to hold it would be enough that $\|z\|$ is of order $r + \|y_0\|$ for all $z \in \partial B(y_0, R)$; this can be seen to be so in at least one of the previous cases (look at the left side of Figure 4.6: if $\partial B(y_0, Cr)$ is "too close" to the origin, then $\partial B(y_0, 2Cr)$ is not).

---

[7] The fact that $\widehat{S}$ is uniformly elliptic is almost evident, but see Exercise 4.10.

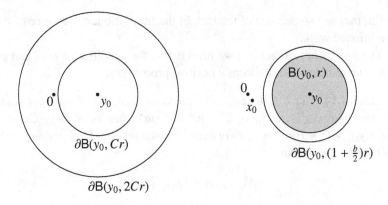

Figure 4.6 On the proof of Lemma 4.14.

For definiteness, assume now that (4.41) holds for $R = 2Cr$. By (4.40), we have then

$$
\begin{aligned}
&\mathbb{P}_{x_0}[\hat{\tau}_{\partial B(y_0,Cr)} < \hat{\tau}_{B(y_0,r)}] \\
&\geq \mathbb{P}_{x_0}[\hat{\tau}_{\partial B(y_0,2Cr)} < \hat{\tau}_{B(y_0,r)}] \\
&\geq \frac{1}{2}\Big(\mathbb{P}_{x_0}[\tau_{\partial B(y_0,2Cr)} < \tau_{B(y_0,r)}] - \mathbb{P}_{x_0}[\tau_0 < \tau_{\partial B(y_0,2Cr)}]\Big) \\
&\geq \frac{1}{2}\Big(c'' - \mathbb{P}_{x_0}[\tau_0 < \tau_{\partial B(y_0,2Cr)}]\Big) \\
&\geq \frac{c''}{4},
\end{aligned}
$$

provided that

$$
\mathbb{P}_{x_0}[\tau_0 < \tau_{\partial B(y_0,2Cr)}] \leq \frac{c''}{2}. \tag{4.42}
$$

Now, if $0 \notin B(y_0, Cr)$, then (4.42) trivially holds; so, let us assume that $0 \in B(y_0, Cr)$. We then consider two cases: $\|x_0\| \geq \frac{b}{4}r$, and $\|x_0\| < \frac{b}{4}r$. In the first case, Lemma 3.12 implies that $\mathbb{P}_{x_0}[\tau_0 < \tau_{\partial B(y_0,2Cr)}] \asymp \frac{1}{\ln r}$, so (4.42) holds for large enough r. In the second case, note first that

$$
\mathbb{P}_{x_0}[\hat{\tau}_{\partial B(y_0,Cr)} < \hat{\tau}_{B(y_0,r)}] \geq \min_{z \in \partial B(y_0,(1+\frac{b}{2})r)} \mathbb{P}_z[\hat{\tau}_{\partial B(y_0,Cr)} < \hat{\tau}_{B(y_0,r)}]
$$

and, for all $z \in \partial B(y_0, (1 + \frac{b}{2})r)$, it holds that $\|z\| \geq \frac{b}{4}r$ (see Figure 4.6 on the right). We may then repeat this reasoning with an arbitrary $z \in \partial B(y_0, (1 + \frac{b}{2})r)$ on the place of x_0 to finally obtain the claim. $\square$

Next technical fact we need is the following lower bound on the probability that the conditioned walk never hits a disk:

Lemma 4.15. *Fix $b > 0$ and assume that $x_0, y_0 \in \mathbb{Z}^2 \setminus \{0\}$ and $r \geq 1$ are such that $\|x_0 - y_0\| \geq (1 + b)r$. Then there exists $c = c(b)$ such that for all x_0, y_0, r as before it holds that*

$$\mathbb{P}_{x_0}[\hat{\tau}_{B(y_0,r)} = \infty] \geq \frac{c}{\ln(\|y_0\| + r)}. \tag{4.43}$$

Proof We need to consider two cases: $B(y_0, r)$ is (*relatively* to its size) close to/far from the origin. First, let us assume that $\|y_0\| < 12r$ (so that the disk is relatively close to the origin). In this case, we can assume additionally that $\|x_0 - y_0\| \geq 51r$ (indeed, Lemma 4.14 implies that, for any starting position x_0 such that $\|x_0 - y_0\| \geq (1 + b)r$, with at least a constant probability the walk reaches $\partial B(y_0, 51r)$ before hitting $B(y_0, r)$). Then it holds that $B(y_0, r) \subset B(13r)$, and[8] $B(26r) \subset B(y_0, 51r)$. Now, Corollary 4.7 easily implies that, if $r \geq 1$ and $\|x\| \geq 2r$,

$$\mathbb{P}_x[\hat{\tau}_{B(r)} = \infty] \geq \frac{c'}{\ln r}$$

(because (4.10) will work for large enough $\|x\|$, and one can use the uniform ellipticity of $\widehat{S}$ otherwise); this proves (4.43) in the first case.

Now suppose that $\|y_0\| \geq 12r$ (that is, $r \leq \frac{1}{12}\|y_0\|$). Analogously to the previous case, Lemma 4.14 permits us to assume without loss of generality that $x_0 \in \partial B(y_0, 3r)$. We now use the martingale (recall Proposition 4.10)

$$\hat{\ell}(\widehat{S}_{n \wedge \hat{\tau}_{y_0}}, y_0) = 1 + \frac{a(y_0) - a(\widehat{S}_{n \wedge \hat{\tau}_{y_0}} - y_0)}{a(\widehat{S}_{n \wedge \hat{\tau}_{y_0}})}.$$

In *exactly*[9] the same way as in the proof of Theorem 2.5 (page 20), we obtain from the optional stopping theorem that

$$\hat{\ell}(x_0, y_0) = \sum_{z \in \partial B(y_0,r)} \mathbb{P}_{x_0}[\hat{\tau}_{B(y_0,r)} < \infty, \widehat{S}_{\hat{\tau}_{B(y_0,r)}} = z]\hat{\ell}(z, y_0)$$

$$\geq \mathbb{P}_{x_0}[\hat{\tau}_{B(y_0,r)} < \infty] \min_{z \in \partial B(y_0,r)} \hat{\ell}(z, y_0),$$

so

$$\mathbb{P}_{x_0}[\hat{\tau}_{B(y_0,r)} < \infty] \leq \frac{\hat{\ell}(x_0, y_0)}{\min_{z \in \partial B(y_0,r)} \hat{\ell}(z, y_0)}. \tag{4.44}$$

---

[8] The reader is advised to make a picture.
[9] Recall also (4.18).

Assume $z \in \partial B(y_0, r)$ and write, using (3.36) and with $\gamma'' := \pi \gamma'/2$,

$$
\begin{aligned}
\hat{\ell}(z, y_0) &= \frac{a(z) + a(y_0) - a(y_0 - z)}{a(z)} \\
&= \frac{\ln \|z\| + \ln \|y_0\| - \ln r + \gamma'' + O(\|y_0\|^{-2} + r^{-1})}{\ln \|z\| + \gamma'' + O(\|z\|^{-2})} \\
&\geq \frac{\ln(\|y_0\| - r) + \ln \|y_0\| - \ln r + \gamma'' + O(\|y_0\|^{-2} + r^{-1})}{\ln(\|y_0\| + r) + \gamma'' + O(\|y_0\|^{-2} + r^{-1})} \\
&= \frac{2 \ln \|y_0\| + \ln\left(1 - \frac{r}{\|y_0\|}\right) - \ln r + \gamma'' + O(r^{-1})}{\ln \|y_0\| + \ln\left(1 + \frac{r}{\|y_0\|}\right) + \gamma'' + O(r^{-1})} \\
&:= \frac{T_1}{T_2},
\end{aligned}
\tag{4.45}
$$

and, denoting $R := \|x_0 - y_0\|$ (so that $R = 3r + O(1)$),

$$
\begin{aligned}
\hat{\ell}(x_0, y_0) &= \frac{a(x_0) + a(y_0) - a(y_0 - x_0)}{a(x_0)} \\
&= \frac{\ln \|x_0\| + \ln \|y_0\| - \ln R + \gamma'' + O(\|y_0\|^{-2} + R^{-2})}{\ln \|x_0\| + \gamma'' + O(\|x_0\|^{-2})} \\
&\leq \frac{\ln(\|y_0\| + R) + \ln \|y_0\| - \ln R + \gamma'' + O(\|y_0\|^{-2} + R^{-2})}{\ln(\|y_0\| - R) + \gamma'' + O(\|y_0\|^{-2} + R^{-2})} \\
&= \frac{2 \ln \|y_0\| + \ln\left(1 + \frac{R}{\|y_0\|}\right) - \ln R + \gamma'' + O(R^{-2})}{\ln \|y_0\| + \ln\left(1 - \frac{R}{\|y_0\|}\right) + \gamma'' + O(R^{-2})} \\
&:= \frac{T_3}{T_4}.
\end{aligned}
\tag{4.46}
$$

A straightforward calculation yields

$$
\frac{T_2}{T_4} = 1 + \frac{\ln \frac{1 + r/\|y_0\|}{1 - R/\|y_0\|} + O(r^{-1})}{\ln \|y_0\| + \ln\left(1 - \frac{R}{\|y_0\|}\right) + \gamma'' + O(R^{-2})},
$$

and

$$
\begin{aligned}
\frac{T_3}{T_1} &= 1 - \frac{\ln\left(\frac{R}{r} \cdot \frac{1 - r/\|y_0\|}{1 + R/\|y_0\|}\right) + O(r^{-1})}{2 \ln \|y_0\| - \ln \frac{r}{1 - r/\|y_0\|} + \gamma'' + O(r^{-1})} \\
&\leq 1 - \frac{\ln\left(\frac{R}{r} \cdot \frac{1 - r/\|y_0\|}{1 + R/\|y_0\|}\right) + O(r^{-1})}{2 \ln \|y_0\| + \gamma''}.
\end{aligned}
$$

Therefore, by (4.44) we have (after some more calculations, sorry)

$$
\mathbb{P}_{x_0}[\hat{\tau}_{B(y_0, r)} < \infty]
$$

$$\leq \frac{T_2}{T_4} \times \frac{T_3}{T_1}$$

$$\leq 1 - \frac{\ln\left(\frac{R}{r} \cdot \frac{1-r/\|y_0\|}{1+R/\|y_0\|}\right)^{1/2} - \ln\frac{1+r/\|y_0\|}{1-R/\|y_0\|} + O((\ln r)^{-1})}{\ln\|y_0\|(1 + O(\frac{1}{\ln\|y_0\|}))}. \qquad (4.47)$$

It remains only to observe that, if r is large enough, the numerator in (4.47) is bounded from below by a positive constant: indeed, observe that $\frac{R}{r}$ is (asymptotically) 3, $\frac{r}{\|y_0\|}$ and $\frac{R}{\|y_0\|}$ are at most $\frac{1}{12}$ and $\frac{1}{4}$ respectively, and

$$\sqrt{3 \times \frac{1-\frac{1}{12}}{1+\frac{1}{4}} \times \frac{1-\frac{1}{4}}{1+\frac{1}{12}}} = \sqrt{\frac{891}{845}} > 1.$$

This concludes the proof of Lemma 4.15 in the case when r is large enough; the case of smaller values of r, though, can be easily reduced to the former one by using the uniform ellipticity of the $\widehat{S}$-walk. $\qquad \square$

Now we profit from the previous technical lemmas to obtain a quantitative result for the entrance measure to a finite set; it is quite analogous to Theorem 3.17, only this time for the conditioned walk:

Theorem 4.16. *Assume that $A \subset \mathbb{Z}^2 \setminus \{0\}$ is finite and $x \notin A$ is such that* $\text{dist}(x, A) \geq 12(\text{diam}(A) + 1)$. *For all $y \in A$, we have*

$$\mathbb{P}_x[\widehat{S}_{\hat{\tau}_A} = y \mid \hat{\tau}_A < \infty] = \widehat{\text{hm}}_A(y)(1 + O(\tfrac{\text{diam}(A)}{\text{dist}(x,A)})). \qquad (4.48)$$

Proof We will begin as in the proof of Theorem 3.8 (page 43), and the reader is also advised to recall the solution of Exercise 3.25, since the next proof will have quite a few similarities to it. Let us assume without restricting generality that A contains at least two sites, so that $\text{diam}(A) \geq 1$. First, we proceed very similarly to the proof of Theorem 3.8. We keep the notation $\Theta_{xy}^{(n)}$ from that proof; also, recall that P_φ (respectively, $\widehat{P}_\varphi$) is the weight of the trajectory φ with respect to the SRW (respectively, to the $\widehat{S}$-walk). We also keep (now with hats, as we are dealing with the $\widehat{S}$-walk) the notations $\widehat{N}_x, \widehat{N}_x^\flat, \widehat{N}_x^\sharp$ for, respectively, the total number of visits to $x \notin A$, the number of visits to x before the first return to A, and the number of visits to x after the first return to A (again, setting $\widehat{N}_x^\sharp = 0$ on $\{\hat{\tau}_A^+ = \infty\}$).

As in (3.22), it is clear that

$$\mathbb{P}_x[\hat{\tau}_A < \infty, \widehat{S}_{\hat{\tau}_A} = y] = \sum_{n=1}^{\infty} \sum_{\varphi \in \Theta_{xy}^{(n)}} \widehat{P}_\varphi$$

but, due to Proposition 4.3 (i), the relation corresponding to (3.23) will now be

$$P_y[\widehat{N}_x^\flat \geq n]\frac{a^2(y)}{a^2(x)} = \sum_{\wp\in\Theta_{xy}^{(n)}} \widehat{P}_\wp.$$

Then, analogously to (3.24), we write

$$P_x[\widehat{S}_{\hat\tau_A} = y \mid \hat\tau_A < \infty]$$

$$= \frac{P_x[\hat\tau_A < \infty, \widehat{S}_{\hat\tau_A} = y]}{P_x[\hat\tau_A < \infty]}$$

$$= \frac{1}{P_x[\hat\tau_A < \infty]} \sum_{n=1}^\infty \frac{a^2(y)}{a^2(x)} P_y[\widehat{N}_x^\flat \geq n]$$

$$= \frac{a^2(y)}{a^2(x)P_x[\hat\tau_A < \infty]} \mathbb{E}_y \widehat{N}_x^\flat$$

$$= \frac{a^2(y)}{a^2(x)P_x[\hat\tau_A < \infty]} (\mathbb{E}_y\widehat{N}_x - \mathbb{E}_y\widehat{N}_x^\sharp)$$

$$= \frac{a^2(y)}{P_x[\hat\tau_A < \infty]}\Big(\frac{\widehat{G}(y,x)}{a^2(x)} - \sum_{z\in\partial A} P_y[\hat\tau_A^+ < \infty, \widehat{S}_{\hat\tau_A^+} = z]\frac{\widehat{G}(z,x)}{a^2(x)}\Big)$$

$$= \frac{a^2(y)}{P_x[\hat\tau_A < \infty]}\Big(\hat{g}(y,x) - \sum_{z\in\partial A} P_y[\hat\tau_A^+ < \infty, \widehat{S}_{\hat\tau_A^+} = z]\hat{g}(z,x)\Big)$$

$$= \frac{a^2(y)}{P_x[\hat\tau_A < \infty]}\Big(\hat{g}(y,x)\Big(\widehat{Es}_A(y) + \sum_{z\in\partial A} P_y[\hat\tau_A^+ < \infty, \widehat{S}_{\hat\tau_A^+} = z]\Big)$$

$$\qquad - \sum_{z\in\partial A} P_y[\hat\tau_A^+ < \infty, \widehat{S}_{\hat\tau_A^+} = z]\hat{g}(z,x)\Big)$$

$$= \frac{a^2(y)\hat{g}(y,x)\widehat{Es}_A(y)}{P_x[\hat\tau_A < \infty]}$$

$$+ \frac{a^2(y)}{P_x[\hat\tau_A < \infty]} \sum_{z\in\partial A} P_y[\hat\tau_A^+ < \infty, \widehat{S}_{\hat\tau_A^+} = z](\hat{g}(y,x) - \hat{g}(z,x)). \qquad (4.49)$$

Next, it is straightforward to obtain that

$$\frac{1}{\hat{g}(x,y)} = O(\ln(1 + \|x\| \vee \|y\|)) \qquad (4.50)$$

(see Exercise 4.14), So, by (4.36) it holds that

$$\frac{a^2(y)\hat{g}(y,x)\widehat{Es}_A(y)}{P_x[\hat\tau_A < \infty]} = \widehat{hm}_A(y)(1 + O(\tfrac{\mathrm{diam}(A)}{\mathrm{dist}(x,A)})). \qquad (4.51)$$

So, it remains to show that the second term in (4.49) is $O(\widehat{hm}_A(y)\frac{\mathrm{diam}(A)}{\mathrm{dist}(x,A)})$.

Similarly to the proof of Theorem 3.8 and the solution of Exercise 3.25, we are going to use the optional stopping theorem with the martingale $\widehat{M}_{n \wedge \hat{\tau}_x}$ where

$$\widehat{M}_n = \hat{g}(y, x) - \hat{g}(\widehat{S}_{n \wedge \hat{\tau}_x}, x)$$

to estimate the second term in (4.49). Recall that $y \in \partial A$, and let us define

$$V = \partial B(y, 2 \operatorname{diam}(A))$$

(the reader is advised to look again at Figure 3.6 on page 61). Let $\tau = \hat{\tau}_A^+ \wedge \hat{\tau}_V$. We have (note that $\tau < \hat{\tau}_x$)

$$\begin{aligned}
0 = \mathbb{E}_y \widehat{M}_0 \\
= \mathbb{E}_y \widehat{M}_\tau \\
= \mathbb{E}_y(\widehat{M}_{\hat{\tau}_A^+} \mathbb{1}\{\hat{\tau}_A^+ < \hat{\tau}_V\}) + \mathbb{E}_y(\widehat{M}_{\hat{\tau}_V} \mathbb{1}\{\hat{\tau}_V < \hat{\tau}_A^+\})
\end{aligned}$$

(since $\mathbb{1}\{\hat{\tau}_A^+ < \infty\} = \mathbb{1}\{\hat{\tau}_A^+ < \hat{\tau}_V\} + \mathbb{1}\{\hat{\tau}_V < \hat{\tau}_A^+ < \infty\}$)

$$\begin{aligned}
= \mathbb{E}_y(\widehat{M}_{\hat{\tau}_A^+} \mathbb{1}\{\hat{\tau}_A^+ < \infty\}) - \mathbb{E}_y(\widehat{M}_{\hat{\tau}_A^+} \mathbb{1}\{\hat{\tau}_V < \hat{\tau}_A^+ < \infty\}) \\
+ \mathbb{E}_y(\widehat{M}_{\hat{\tau}_V} \mathbb{1}\{\hat{\tau}_V < \hat{\tau}_A^+\}).
\end{aligned}$$

Note that for any $z \in V \cup \partial A$ it holds that $\|y - z\| \le 2 \operatorname{diam}(A)$ and $\|x - z\| \ge 10(\operatorname{diam}(A) + 1)$, so in the following we will be able to apply Lemma 4.13. Since

$$\mathbb{E}_y(\widehat{M}_{\hat{\tau}_A^+} \mathbb{1}\{\hat{\tau}_A^+ < \infty\}) = \sum_{z \in \partial A} \mathbb{P}_y[\hat{\tau}_A^+ < \infty, \widehat{S}_{\hat{\tau}_A^+} = z](\hat{g}(y, x) - \hat{g}(z, x)),$$

we obtain that

$$\begin{aligned}
\sum_{z \in \partial A} \mathbb{P}_y[\hat{\tau}_A^+ < \infty, \widehat{S}_{\hat{\tau}_A^+} = z](\hat{g}(y, x) - \hat{g}(z, x)) \\
= \mathbb{E}_y(\widehat{M}_{\hat{\tau}_A^+} \mathbb{1}\{\hat{\tau}_V < \hat{\tau}_A^+ < \infty\}) - \mathbb{E}_y(\widehat{M}_{\hat{\tau}_V} \mathbb{1}\{\hat{\tau}_V < \hat{\tau}_A^+\}) \\
= \mathbb{P}_y[\hat{\tau}_V < \hat{\tau}_A^+]\big(\mathbb{E}_y(\widehat{M}_{\hat{\tau}_A^+} \mathbb{1}\{\hat{\tau}_A^+ < \infty\} \mid \hat{\tau}_V < \hat{\tau}_A^+) - \mathbb{E}_y(\widehat{M}_{\hat{\tau}_V} \mid \hat{\tau}_V < \hat{\tau}_A^+)\big)
\end{aligned}$$

(recall Lemma 4.13 and (4.36); note that $\widehat{M}_\tau = \hat{g}(y, x) - \hat{g}(z, x)$ on $\{\widehat{S}_\tau = z\}$)

$$\le \mathbb{P}_y[\hat{\tau}_V < \hat{\tau}_A^+] \times O\left(\frac{\operatorname{diam}(A)}{\operatorname{dist}(x, A) \ln(1 + \|y\| + \operatorname{diam}(A)) \ln(1 + \|x\| \vee (\|y\| + \operatorname{diam}(A)))}\right). \quad (4.52)$$

Next, we can write

$$\begin{aligned}
\widehat{\mathrm{Es}}_A(y) = \mathbb{P}_y[\hat{\tau}_A^+ = \infty] \\
= \sum_{v \in V} \mathbb{P}_y[\hat{\tau}_V < \hat{\tau}_A^+, \widehat{S}_{\hat{\tau}_V} = v] \mathbb{P}_v[\hat{\tau}_A = \infty]
\end{aligned}$$

(by Lemma 4.15)

$$\geq \frac{c}{\ln(\|y\| + \mathrm{diam}(A))} \mathbb{P}_y[\hat{\tau}_V < \hat{\tau}_A^+],$$

which means that

$$\mathbb{P}_y[\hat{\tau}_V < \hat{\tau}_A^+] \leq O\big(\widehat{\mathrm{Es}}_A(y)\ln(\|y\| + \mathrm{diam}(A))\big). \tag{4.53}$$

Also, (4.51) implies that

$$\frac{a^2(y)}{\mathbb{P}_x[\hat{\tau}_A < \infty]} = O\Big(\frac{\widehat{\mathrm{hm}}_A(y)}{\widehat{\mathrm{Es}}_A(y)\hat{g}(x,y)}\Big). \tag{4.54}$$

It only remains to combine (4.52) through (4.54) with (4.50) to see that the second term in (4.49) is indeed $O\big(\widehat{\mathrm{hm}}_A(y)\frac{\mathrm{diam}(A)}{\mathrm{dist}(x,A)}\big)$, thus concluding the proof of Theorem 4.16. □

In the remaining section of this chapter, we are going to discuss some even more surprising properties of the conditioned walk $\widehat{S}$.

4.5 Range of the conditioned SRW

For a set $T \subset \mathbb{Z}_+$ (thought of as a set of time moments) let

$$\widehat{S}_T = \bigcup_{m \in T} \{\widehat{S}_m\}$$

be the *range* of the walk $\widehat{S}$ with respect to that set, i.e., it is made of sites that are visited by $\widehat{S}$ over T. For simplicity, we assume in the following that the walk $\widehat{S}$ starts at a fixed neighbour x_0 of the origin, and we write $\mathbb{P}$ for $\mathbb{P}_{x_0}$. For a nonempty and finite set $A \subset \mathbb{Z}^2$, let us consider the random variable

$$\mathcal{R}(A) = \frac{|A \cap \widehat{S}_{[0,\infty)}|}{|A|};$$

that is, $\mathcal{R}(A)$ is the proportion of visited sites of A by the walk $\widehat{S}$ (and, therefore, $1 - \mathcal{R}(A)$ is the proportion of unvisited sites of A). In this section, we are interested in the following question: how should $\mathcal{R}(A)$ behave for "large" sets? By (4.21), in average approximately half of A should be covered, i.e., $\mathbb{E}\mathcal{R}(A)$ should be close to $1/2$. Now, keeping in mind how difficult is to find something *really* new in these modern times, let us think what are the usual examples of random variables which are concentrated on $[0, 1]$ and have expected value $1/2$. Well, three examples come to mind: Uniform$[0, 1]$, Bernoulli$(1/2)$, and $1/2$ itself. Which one of them shall the

walk $\widehat{S}$ choose? It turns out that it is the first one which is the most relevant for the conditioned walk.[10] Indeed, the main result of this section is
the following surprising fact: the proportion of visited sites of a "typical"
large set (e.g., a disk, a rectangle, a segment) is a *random variable* which is
close in distribution to the Uniform[0, 1] law. The paper [48] contains the
corresponding results in greater generality, but here we content ourselves
in proving the result for a particular case of a large disk which does not
"touch" the origin:

Theorem 4.17. *Let $D \subset \mathbb{R}^2$ be a closed disk such that $0 \notin D$, and denote
$D_n = nD \cap \mathbb{Z}^2$. Then, for all $s \in [0, 1]$, we have, with positive constant c_1
depending only on D,*

$$\left| \mathbb{P}[\mathcal{R}(D_n) \leq s] - s \right| \leq c_1 \left(\frac{\ln \ln n}{\ln n} \right)^{1/3}. \tag{4.55}$$

Also, we prove that the range of $\widehat{S}$ contains many "big holes".

Theorem 4.18. *Let D and D_n be as in Theorem 4.17. Then it holds that*

$$\mathbb{P}[D_n \cap \widehat{S}_{[0,\infty)} = \emptyset \text{ for infinitely many } n] = 1. \tag{4.56}$$

Theorem 4.18 invites the following:

Remark 4.19. A natural question to ask is whether there are also large disks
that are *completely filled*, that is, if a.s. there are infinitely many n such that
$D_n \subset \widehat{S}_{[0,\infty)}$. It is not difficult to see that the answer to this question is "no".
We, however, postpone explaining that to Chapter 6 (see Exercise 6.10).

We now obtain some refined bounds on the hitting probabilities for excursions of the conditioned walk.

Lemma 4.20. *Let us assume that $x, y \in \mathbb{Z}^2 \setminus \{0\}$ with $x \neq y$, and $R >
\max\{\|x\|, \|y\|\} + 1$. Then we have*

$$\mathbb{P}_x[\hat{\tau}_y < \hat{\tau}_{\partial B(R)}]$$
$$= \frac{a(R)(a(x) + a(y) - a(x - y)) - a(x)a(y)(1 + O(\frac{\|y\|}{R \ln(\|y\|+1)}))}{a(x)(2a(R) - a(y)(1 + O(\frac{\|y\|}{R \ln(\|y\|+1)})))}. \tag{4.57}$$

Proof Once again in this book, we use the optional stopping theorem, but

---

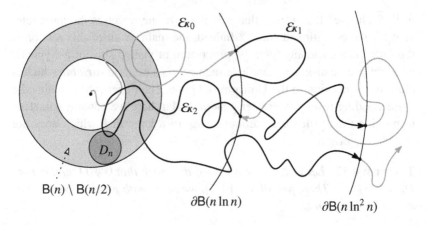

Figure 4.7 Excursions that may visit D_n.

this time with a new martingale, namely, $M_n = \hat{\ell}(\widehat{S}_{n \wedge \hat{\tau}_y}, y)$ (recall Proposition 4.10). We have (recall (4.25))

$$\hat{\ell}(z, y) = \frac{1}{a(z)}(a(y) + a(z) - a(z - y))$$

$$= \frac{1}{a(R)}(1 + O(\tfrac{1}{R \ln R})) \times (a(y) + O(\tfrac{\|y\|}{R}))$$

$$= \frac{a(y)}{a(R)}(1 + O(\tfrac{\|y\|}{R \ln(\|y\|+1)})) \qquad (4.58)$$

for any $z \in \partial B(R)$. Let us then abbreviate $p = \mathbb{P}_x[\hat{\tau}_y < \hat{\tau}_{\partial B(R)}]$ and write, using the aforementioned theorem with stopping time $\hat{\tau}_y \wedge \hat{\tau}_{\partial B(R)}$

$$\mathbb{E}_x M_0 = \frac{1}{a(x)}(a(x) + a(y) - a(x - y))$$

(note that $\hat{\ell}(y, y) = 2$)

$$= 2p + (1 - p)\mathbb{E}_x(\hat{\ell}(\widehat{S}_{\hat{\tau}_{\partial B(R)}}, y) \mid \hat{\tau}_{\partial B(R)} < \hat{\tau}_y)$$

(by (4.58))

$$= 2p + (1 - p)\frac{a(y)}{a(R)}(1 + O(\tfrac{\|y\|}{R \ln(\|y\|+1)})).$$

Solving this equation for p, we obtain (4.57). ☐

Let us assume that $x \in B(n \ln n)$, $y \in B(n)$, and abbreviate $R = n \ln^2 n$. A quick calculation using Lemma 4.20 then shows that

$$\mathbb{P}_x[\hat{\tau}_y < \hat{\tau}_{\partial B(R)}]$$

$$= \frac{a(R)(a(x) + a(y) - a(x-y)) - a(x)a(y)(1 + O(\ln^{-3} n))}{a(x)(2a(R) - a(y)(1 + O(\ln^{-3} n)))}. \qquad (4.59)$$

We proceed with

Proof of Theorem 4.17 First, we describe informally the idea of the proof. We consider the visits to the set D_n during excursions of the walk from $\partial B(n \ln n)$ to $\partial B(n \ln^2 n)$; see Figure 4.7. The crucial argument is the following: the randomness of $\mathcal{R}(D_n)$ comes from the *number* Q of those excursions and not from the excursions themselves. If the number of excursions is approximately $c \times \frac{\ln n}{\ln \ln n}$, then it is possible to show, using a standard weak law of large numbers (LLN) argument, that the proportion of covered sites in D_n is concentrated around $1 - e^{-c}$. On the other hand, that number of excursions can be modeled roughly as $Y \times \frac{\ln n}{\ln \ln n}$, where Y is an exponential random variable with rate 1. We then find that

$$\mathbb{P}[\mathcal{R}(D_n) \geq 1 - s] \approx \mathbb{P}[Y \geq \ln s^{-1}] = s,$$

as required.

In the following, we will assume, for concreteness, that $B(1/2) \subset D \subset B(1)$ so that $D_n \subset B(n) \setminus B(n/2)$ for all n; the extension of the proof to the general case is straightforward.

We now give a rigorous argument. Let $\widehat{H}$ be the conditional entrance measure for the (conditioned) walk $\widehat{S}$, i.e.,

$$\widehat{H}_{D_n}(x, y) = \mathbb{P}_x[\widehat{S}_{\hat{\tau}_{D_n}} = y \mid \hat{\tau}_{D_n} < \infty]. \qquad (4.60)$$

Let us denote the initial piece of the trajectory by $\mathcal{E}_0 = \widehat{S}_{[0, \hat{\tau}_{\partial B(n \ln n)}]}$. Then, we consider a *Markov chain* $(\mathcal{E}_k, k \geq 1)$ of excursions between $\partial B(n \ln n)$ and $\partial B(n \ln^2 n)$, defined in the following way: for $k \geq 2$, the initial site of $\mathcal{E}_k$ is chosen according to the measure $\widehat{H}_{B(n \ln n)}(z_{k-1}, \cdot)$, where $z_{k-1} \in \partial B(n \ln^2 n)$ is the last site of the excursion $\mathcal{E}_{k-1}$; also, the initial site of $\mathcal{E}_1$ is the last site of $\mathcal{E}_0$; the weights of trajectories are chosen according to (4.5) (i.e., each excursion is an $\widehat{S}$-walk trajectory). It is important to observe that one may couple $(\mathcal{E}_k, k \geq 1)$ with the "true" excursions of the walk $\widehat{S}$ in an obvious way: one just picks the excursions subsequently, each time tossing a coin to decide if the walk returns to $B(n \ln n)$.

Let

$$\psi_n = \min_{x \in \partial B(n \ln^2 n)} \mathbb{P}_x[\hat{\tau}_{\partial B(n \ln n)} = \infty],$$

$$\psi_n^* = \max_{x \in \partial B(n \ln^2 n)} \mathbb{P}_x[\hat{\tau}_{\partial B(n \ln n)} = \infty]$$

be the minimal and maximal probabilities to avoid $B(n \ln n)$, starting at sites of $\partial B(n \ln^2 n)$. Using (4.10) together with (3.36), we obtain

$$
\mathbb{P}_x[\hat{\tau}_{\partial B(n \ln n)} = \infty] = 1 - \frac{a(n \ln n) + O((n \ln n)^{-1})}{a(x)}
$$

$$
= \frac{a(n \ln^2 n) - a(n \ln n) + O((n \ln n)^{-1})}{a(n \ln^2 n) + O((n \ln^2 n)^{-1})}
$$

(by (3.36)–(3.37), with $\gamma'' = \frac{\pi \gamma'}{2} = \gamma + \frac{3}{2} \ln 2$)

$$
= \frac{\ln \ln n}{\ln n + 2 \ln \ln n + \gamma''} (1 + o(n^{-1})) \qquad (4.61)
$$

for any $x \in \partial B(n \ln^2 n)$, and so it also holds that

$$
\psi_n = \frac{\ln \ln n}{\ln n + 2 \ln \ln n + \gamma''} (1 + o(n^{-1})). \qquad (4.62)
$$

Let us consider a sequence of i.i.d. random variables $(\eta_k, k \geq 0)$ such that $\mathbb{P}[\eta_k = 1] = 1 - \mathbb{P}[\eta_k = 0] = \psi_n$. Let $\widehat{Q} = \min\{k : \eta_k = 1\}$, so that $\widehat{Q}$ is a geometric random variable with mean ψ_n^{-1}.

Lemma 4.21. *The random variable $\widehat{Q}$ can be coupled with the actual number of excursions Q in such a way that $Q \leq \widehat{Q}$ a.s. and*

$$
\mathbb{P}[Q \neq \widehat{Q}] \leq o(n^{-1}). \qquad (4.63)
$$

Moreover, this coupling can be constructed in such a way that $\widehat{Q}$ is independent from the excursion sequence $(\mathcal{E}_k, k \geq 1)$ itself.

Proof Let $U_1, U_2, U_3, \dots$ be a sequence of i.i.d. random variables with Uniform$[0, 1]$ distribution; we will now explain how to construct all random variables of interest using this sequence. First, we set

$$
\eta_k = \mathbf{1}\{U_k \leq \psi_n\}.
$$

Next, let $\tilde{\eta}_k = \mathbf{1}\{Q < k\}$ be the indicator of the event that the walk $\widehat{S}$ does not make its kth excursion. Given that $\tilde{\eta}_{k-1} = 0$ and z_{k-1} is the last site of $\mathcal{E}_{k-1}$, set

$$
\tilde{\eta}_k = \mathbf{1}\{U_k \leq \mathbb{P}_{z_{k-1}}[\hat{\tau}_{\partial B(n \ln n)} = \infty]\}.
$$

It is clear that, with this construction, $Q \leq \widehat{Q}$ and that $\widehat{Q}$ is independent of the excursions themselves. Then, we write

$$
\{Q > \widehat{Q}\} = \{\text{there exists } k \leq \widehat{Q} - 1
$$
$$
\text{such that } \psi_n < U_k \leq \mathbb{P}_{z_{k-1}}[\hat{\tau}_{\partial B(n \ln n)} = \infty]\},
$$

so

$$\mathbb{P}[Q > \widehat{Q}] \leq \mathbb{P}[\text{there exists } k \leq \widehat{Q} - 1 \text{ such that } \psi_n < U_k \leq \psi_n^*]. \quad (4.64)$$

In the following, assume without loss of generality that n is large enough so that $\psi_n < 1/2$. By (4.64), we have

$$\mathbb{P}[Q < k \mid \widehat{Q} = k]$$
$$\leq \mathbb{P}[\text{there exists } j \leq k - 1 \text{ such that } \psi_n < U_k \leq \psi_n^*$$
$$\mid U_j > \psi_n \text{ for } j = 1, \ldots, k - 1, U_k \leq \varphi_n]$$
$$\leq 2(\psi_n^* - \psi_n)k.$$

The expression (4.61) implies that $\psi_n^* - \psi_n \leq o(\frac{\ln \ln n}{n \ln n})$ for any $x \in \partial B(n \ln^2 n)$, and so (recall that, by (4.62), $1/\psi_n = O(\frac{\ln n}{\ln \ln n})$)

$$\mathbb{P}[Q < \widehat{Q}] \leq \sum_{k=1}^{\infty} (1 - \varphi_n)^{k-1} \psi_n \times 2(\psi_n^* - \psi_n)k$$
$$= \frac{2(\psi_n^* - \psi_n)}{\psi_n} = o(n),$$

which concludes the proof. □

We continue proving Theorem 4.17. Define

$$\mathcal{R}^{(k)} = \frac{\left| D_n \cap (\mathcal{E}_0 \cup \mathcal{E}_1 \cup \ldots \cup \mathcal{E}_k) \right|}{|D_n|},$$

to be the proportion of visited sites in D_n with respect to the first k excursions together with the initial piece $\mathcal{E}_0$. According to the programme outlined in the beginning of the proof, we now aim to verify that that $\mathcal{R}^{(k)}$ is concentrated around its mean value; we do this using a typical weak-law-of-large-numbers argument.

Now it is straightforward to check[11] that (4.59) implies that, for any $x \in \partial B(n \ln n)$ and $y \in D_n$,

$$\mathbb{P}_x[\hat{\tau}_y < \hat{\tau}_{\partial B(n \ln^2 n)}] = \frac{\ln \ln n}{\ln n}\left(1 + O(\frac{\ln \ln n}{\ln n})\right), \quad (4.65)$$

and, for $y, z \in B(n) \setminus B(n/2)$ such that $\|y - z\| = n/b$ with $b \leq 2 \ln n$,

$$\mathbb{P}_z[\hat{\tau}_y < \hat{\tau}_{\partial B(n \ln^2 n)}] = \frac{2 \ln \ln n + \ln b}{\ln n}\left(1 + O(\frac{\ln \ln n}{\ln n})\right) \quad (4.66)$$

(we leave checking this to the reader, as Exercise 4.20).

---

[11] The calculation is a bit long, though; consistent with the approach of this book, it is omitted.

Next, for $y \in D_n$ and a fixed $k \geq 1$, consider the random variable

$$\xi_y^{(k)} = \mathbb{1}\{y \notin \mathcal{E}_0 \cup \mathcal{E}_1 \cup \ldots \cup \mathcal{E}_k\},$$

so that $1 - \mathcal{R}^{(k)} = |D_n|^{-1} \sum_{y \in D_n} \xi_y^{(k)}$. Now (4.65) implies that, for all $j \geq 1$,

$$\mathbb{P}[y \notin \mathcal{E}_j] = 1 - \frac{\ln \ln n}{\ln n}\big(1 + O(\tfrac{\ln \ln n}{\ln n})\big),$$

and (4.66) implies that

$$\mathbb{P}[y \notin \mathcal{E}_0] = 1 - O(\tfrac{\ln \ln n}{\ln n})$$

for any $y \in D_n$. Let $\mu_y^{(k)} = \mathbb{E}\xi_y^{(k)}$. Then we have

$$\mu_y^{(k)} = \mathbb{P}[y \notin \mathcal{E}_0 \cup \mathcal{E}_1 \cup \ldots \cup \mathcal{E}_k]$$

$$= \big(1 - O(\tfrac{\ln \ln n}{\ln n})\big) \times \Big(\big(1 - \frac{\ln \ln n}{\ln n}\big(1 + O(\tfrac{\ln \ln n}{\ln n})\big)\big)\Big)^k$$

$$= \exp\Big(-k\frac{\ln \ln n}{\ln n}\big(1 + O(k^{-1} + \tfrac{\ln \ln n}{\ln n})\big)\Big). \tag{4.67}$$

Next, we need to estimate the covariance[12] of $\xi_y^{(k)}$ and $\xi_z^{(k)}$ in case $\|y - z\| \geq \frac{n}{\ln n}$. First note that, for any $x \in \partial B(n \ln n)$,

$$\mathbb{P}_x[\{y, z\} \cap \mathcal{E}_1 = \emptyset] = 1 - \mathbb{P}_x[y \in \mathcal{E}_1] - \mathbb{P}_x[z \in \mathcal{E}_1] + \mathbb{P}_x[\{y, z\} \subset \mathcal{E}_1]$$

$$= 1 - 2\frac{\ln \ln n}{\ln n}\big(1 + O(\tfrac{\ln \ln n}{\ln n})\big) + \mathbb{P}_x[\{y, z\} \subset \mathcal{E}_1]$$

by (4.65); also, since

$$\{\hat{\tau}_y < \hat{\tau}_z < \hat{\tau}_{\partial B(n \ln^2 n)}\} \subset \{\hat{\tau}_y < \hat{\tau}_{\partial B(n \ln^2 n)},$$

$$\widehat{S}_k = z \text{ for some } \hat{\tau}_y < k < \hat{\tau}_{\partial B(n \ln^2 n)}\},$$

from (4.65) and (4.66) we obtain

$$\mathbb{P}_x[\{y, z\} \subset \mathcal{E}_1] = \mathbb{P}_x[\max\{\hat{\tau}_y, \hat{\tau}_z\} < \hat{\tau}_{\partial B(n \ln^2 n)}]$$

$$= \mathbb{P}_x[\hat{\tau}_y < \hat{\tau}_z < \hat{\tau}_{\partial B(n \ln^2 n)}] + \mathbb{P}_x[\hat{\tau}_z < \hat{\tau}_y < \hat{\tau}_{\partial B(n \ln^2 n)}]$$

$$\leq \mathbb{P}_x[\hat{\tau}_y < \hat{\tau}_{\partial B(n \ln^2 n)}]\mathbb{P}_y[\hat{\tau}_z < \hat{\tau}_{\partial B(n \ln^2 n)}]$$

$$\quad + \mathbb{P}_x[\hat{\tau}_z < \hat{\tau}_{\partial B(n \ln^2 n)}]\mathbb{P}_z[\hat{\tau}_y < \hat{\tau}_{\partial B(n \ln^2 n)}]$$

$$\leq 2\frac{\ln \ln n}{\ln n} \times \frac{3 \ln \ln n}{\ln n}\big(1 + O(\tfrac{\ln \ln n}{\ln n})\big)$$

$$= O\big(\big(\frac{\ln \ln n}{\ln n}\big)^2\big).$$

---

[12] Recall that we want a weak-LLN-type argument, that is, we need to estimate $\mathrm{Var} \sum_{y \in D_n} \xi_y^{(k)}$; so, we will indeed need these covariances.

Therefore, similarly to (4.67) we obtain

$$\mathbb{E}(\xi_y^{(k)}\xi_z^{(k)}) = \exp\left(-2k\frac{\ln\ln n}{\ln n}\left(1 + O\left(k^{-1} + \tfrac{\ln\ln n}{\ln n}\right)\right)\right),$$

which, together with (4.67), implies after some elementary calculations that, for all $y, z \in D_n$ such that $\|y - z\| \geq \frac{n}{\ln n}$

$$\text{cov}(\xi_y^{(k)}, \xi_z^{(k)}) = O\left(\tfrac{\ln\ln n}{\ln n}\right) \tag{4.68}$$

uniformly in k, since

$$\left(\frac{\ln\ln n}{\ln n} + k\left(\frac{\ln\ln n}{\ln n}\right)^2\right)\exp\left(-2k\frac{\ln\ln n}{\ln n}\right) = O\left(\tfrac{\ln\ln n}{\ln n}\right)$$

uniformly in k. At this point, we are ready to write that weak-LLN argument. Using Chebyshev's inequality, we write

$$\mathbb{P}\left[\left||D_n|^{-1}\sum_{y\in D_n}(\xi_y^{(k)} - \mu_y^{(k)})\right| > \varepsilon\right]$$

$$\leq (\varepsilon|D_n|)^{-2}\text{Var}\left(\sum_{y\in D_n}\xi_y^{(k)}\right)$$

$$= (\varepsilon|D_n|)^{-2}\sum_{y,z\in D_n}\text{cov}(\xi_y^{(k)}, \xi_z^{(k)})$$

$$= (\varepsilon|D_n|)^{-2}\left(\sum_{\substack{y,z\in D_n, \\ \|y-z\|<\frac{n}{\ln n}}}\text{cov}(\xi_y^{(k)}, \xi_z^{(k)}) + \sum_{\substack{y,z\in D_n, \\ \|y-z\|\geq\frac{n}{\ln n}}}\text{cov}(\xi_y^{(k)}, \xi_z^{(k)})\right)$$

$$\leq (\varepsilon|D_n|)^{-2}\left(\sum_{y\in D_n}\left|D_n\cap B(y, \tfrac{n}{\ln n})\right| + |D_n|^2 O\left(\tfrac{\ln\ln n}{\ln n}\right)\right)$$

$$\leq \varepsilon^{-2}O(\ln^{-2}n + \tfrac{\ln\ln n}{\ln n}) = \varepsilon^{-2}O\left(\tfrac{\ln\ln n}{\ln n}\right). \tag{4.69}$$

Now, having established that $\mathcal{R}^{(k)}$ is "almost deterministic", we recall that the number of excursions is "almost geometrically distributed", and work with this fact. Let

$$\Phi^{(s)} = \min\{k : \mathcal{R}^{(k)} \geq 1 - s\}$$

be the number of excursions necessary to make the unvisited proportion of D_n at most s. We have

$$\mathbb{P}[\mathcal{R}(D_n) \geq 1 - s]$$

$$= \mathbb{P}[\Phi^{(s)} \leq Q]$$

$$= \mathbb{P}[\Phi^{(s)} \leq Q, Q = \widehat{Q}] + \mathbb{P}[\Phi^{(s)} \leq Q, Q \neq \widehat{Q}]$$

$$= \mathbb{P}[\Phi^{(s)} \leq \widehat{Q}] + \mathbb{P}[\Phi^{(s)} \leq Q, Q \neq \widehat{Q}] - \mathbb{P}[\Phi^{(s)} \leq \widehat{Q}, Q \neq \widehat{Q}],$$

so, recalling (4.63),

$$\left|\mathbb{P}[\mathcal{R}(D_n) \geq 1 - s] - \mathbb{P}[\Phi^{(s)} \leq \widehat{Q}]\right| \leq \mathbb{P}[Q \neq \widehat{Q}] \leq O(n^{-1}). \qquad (4.70)$$

Next, by the "independence" part of Lemma 4.21, we write

$$\mathbb{P}[\Phi^{(s)} \leq \widehat{Q}] = \mathbb{E}(\mathbb{P}[\widehat{Q} \geq \Phi^{(s)} \mid \Phi^{(s)}])$$
$$= \mathbb{E}(1 - \psi_n)^{\Phi^{(s)}}, \qquad (4.71)$$

and focus on obtaining lower and upper bounds on the expectation in the right-hand side of (4.71). The idea is that, since $\mathcal{R}^{(k)}$ is concentrated, $\Phi^{(s)}$ has to be concentrated as well; however, since it is in the exponent, some extra care needs to be taken. Let us assume that $s \in (0, 1)$ is fixed and abbreviate

$$\delta_n = \left(\frac{\ln\ln n}{\ln n}\right)^{1/3}$$

$$k_n^- = \left\lfloor (1 - \delta_n) \ln s^{-1} \frac{\ln n}{\ln\ln n} \right\rfloor,$$

$$k_n^+ = \left\lceil (1 + \delta_n) \ln s^{-1} \frac{\ln n}{\ln\ln n} \right\rceil;$$

we also assume that n is sufficiently large so that $\delta_n \in (0, \frac{1}{2})$ and $1 < k_n^- < k_n^+$. Now, according to (4.67),

$$\mu_y^{(k_n^\pm)} = \exp\left(-(1 \pm \delta_n) \ln s^{-1}\left(1 + O((k_n^\pm)^{-1} + \tfrac{\ln\ln n}{\ln n})\right)\right)$$
$$= s \exp\left(-\ln s^{-1}\left(\pm\delta_n + O((k_n^\pm)^{-1} + \tfrac{\ln\ln n}{\ln n})\right)\right)$$
$$= s\left(1 + O(\delta_n \ln s^{-1} + \tfrac{\ln\ln n}{\ln n}(1 + \ln s^{-1}))\right),$$

so in both cases it holds that (observe that $s \ln s^{-1} \leq 1/e$ for all $s \in [0, 1]$)

$$\mu_y^{(k_n^\pm)} = s + O(\delta_n + \tfrac{\ln\ln n}{\ln n}) = s + O(\delta_n). \qquad (4.72)$$

With a similar calculation, one can also obtain that

$$(1 - \psi_n)^{(k_n^\pm)} = s + O(\delta_n). \qquad (4.73)$$

We then write, using (4.72),

$$\mathbb{P}[\Phi^{(s)} > k_n^+] = \mathbb{P}[\mathcal{R}^{(k_n^+)} < 1 - s]$$
$$= \mathbb{P}\left[|D_n|^{-1} \sum_{y \in D_n} \xi_y^{(k_n^+)} > s\right]$$
$$= \mathbb{P}\left[|D_n|^{-1} \sum_{y \in D_n} (\xi_y^{(k_n^+)} - \mu_y^{(k_n^+)}) > s - |D_n|^{-1} \sum_{y \in D_n} \mu_y^{(k_n^+)}\right]$$

$$= \mathbb{P}\left[|D_n|^{-1} \sum_{y \in D_n} (\xi_y^{(k_n^+)} - \mu_y^{(k_n^+)}) > O(\delta_n)\right]. \tag{4.74}$$

Then (4.69) implies that

$$\mathbb{P}[\Phi^{(s)} > k_n^+] \le O\left(\left(\tfrac{\ln \ln n}{\ln n}\right)^{1/3}\right). \tag{4.75}$$

Quite analogously, one can also obtain that

$$\mathbb{P}[\Phi^{(s)} < k_n^-] \le O\left(\left(\tfrac{\ln \ln n}{\ln n}\right)^{1/3}\right). \tag{4.76}$$

Using (4.73) and (4.75), we then write

$$\begin{aligned}
\mathbb{E}(1 - \psi_n)^{\Phi^{(s)}} &\ge \mathbb{E}\left((1 - \psi_n)^{\Phi^{(s)}} \mathbf{1}\{\Phi^{(s)} \le k_n^+\}\right) \\
&\ge (1 - \psi_n)^{k_n^+} \mathbb{P}[\Phi^{(s)} \le k_n^+] \\
&\ge \left(s - O\left(\left(\tfrac{\ln \ln n}{\ln n}\right)^{1/3}\right)\right)\left(1 - O\left(\left(\tfrac{\ln \ln n}{\ln n}\right)^{1/3}\right)\right),
\end{aligned} \tag{4.77}$$

and, using (4.73) and (4.76),

$$\begin{aligned}
\mathbb{E}(1 - \psi_n)^{\Phi^{(s)}} &= \mathbb{E}\left((1 - \psi_n)^{\Phi^{(s)}} \mathbf{1}\{\Phi^{(s)} \ge k_n^-\}\right) \\
&\quad + \mathbb{E}\left((1 - \psi_n)^{\Phi^{(s)}} \mathbf{1}\{\Phi^{(s)} < k_n^-\}\right) \\
&\le (1 - \psi_n)^{k_n^-} + \mathbb{P}[\Phi^{(s)} < k_n^-] \\
&\le \left(s + O\left(\left(\tfrac{\ln \ln n}{\ln n}\right)^{1/3}\right)\right) + O\left(\left(\tfrac{\ln \ln n}{\ln n}\right)^{1/3}\right).
\end{aligned} \tag{4.78}$$

Therefore, recalling (4.70) and (4.71), we obtain (4.55); this concludes the proof of Theorem 4.17. □

Now we prove that there are "big holes" in the range of $\widehat{S}$:

Proof of Theorem 4.18   For the sake of simplicity, let us still assume that $D \subset B(1) \setminus B(1/2)$; the general case can be treated in a completely analogous way.

Consider two sequences of events

$$E_n = \{\hat{\tau}_{D_{2^{3n-1}}} > \hat{\tau}_{\partial B(2^{3n})}, \|\widehat{S}_j\| > 2^{3n-1} \text{ for all } j \ge \hat{\tau}_{\partial B(2^{3n})}\},$$

$$E_n' = \{\|\widehat{S}_j\| > 2^{3n-1} \text{ for all } j \ge \hat{\tau}_{\partial B(2^{3n})}\}$$

and note that $E_n \subset E_n'$ and $2^{3n-1}D \cap \widehat{S}_{[0,\infty)} = \emptyset$ on E_n. Our goal is to show that, almost surely, an infinite number of events $(E_n, n \ge 1)$ occurs. Observe, however, that the events in each of the preceding two sequences are *not* independent, so the "basic" second Borel–Cantelli lemma will not work.

In the following, we use a generalization of the second Borel–Cantelli lemma, known as the Kochen–Stone theorem [55]: it holds that

$$\mathbb{P}\Big[\sum_{k=1}^{\infty} 1\{E_k\} = \infty\Big] \geq \limsup_{k \to \infty} \frac{(\sum_{i=1}^{k} \mathbb{P}[E_i])^2}{\sum_{i,j=1}^{k} \mathbb{P}[E_i \cap E_j]}. \tag{4.79}$$

We will now prove that there exists a positive constant c_4 such that

$$\mathbb{P}[E_n] \geq \frac{c_4}{n} \quad \text{for all } n \geq 1. \tag{4.80}$$

It is elementary to obtain[13] that, for some $c_5 > 0$,

$$\mathbb{P}_x[\tau_{D_{2^{3n-1}}} > \tau_{\partial B(2^{3n})}, \tau_0^+ > \tau_{\partial B(2^{3n})}] > c_5$$

for all $x \in \partial B(2^{3(n-1)})$. Lemma 4.4 then implies that, for some $c_6 > 0$,

$$\begin{aligned}
\mathbb{P}_x[\hat{\tau}_{D_{2^{3n-1}}} &> \hat{\tau}_{\partial B(2^{3n})}] \\
&= (1 + o(2^{-3n}))\mathbb{P}_x[\tau_{D_{2^{3n-1}}} > \tau_{\partial B(2^{3n})} \mid \tau_0^+ > \tau_{\partial B(2^{3n})}] \\
&= (1 + o(2^{-3n}))\mathbb{P}_x[\tau_{D_{2^{3n-1}}} > \tau_{\partial B(2^{3n})}, \tau_0^+ > \tau_{\partial B(2^{3n})}] > c_6
\end{aligned} \tag{4.81}$$

for all $x \in \partial B(2^{3(n-1)})$. Let us denote, recalling (3.36), $\gamma^* = \frac{\pi}{2} \times \frac{1}{\ln 2} \times \gamma' = \frac{2\gamma + 3\ln 2}{2\ln 2}$. Using (4.10), we then obtain

$$\begin{aligned}
\mathbb{P}_z[\|\widehat{S}_j\| > 2^{3n-1} \text{ for all } j \geq 0] &= 1 - \frac{a(2^{3n-1}) + O(2^{-3n})}{a(2^{3n}) + O(2^{-3n})} \\
&= 1 - \frac{\frac{2}{\pi}(3n-1)\ln 2 + \gamma' + O(2^{-3n})}{\frac{2}{\pi}(3n)\ln 2 + \gamma' + O(2^{-3n})} \\
&= \frac{1}{3n + \gamma^*}(1 + o(2^{-3n})).
\end{aligned} \tag{4.82}$$

for any $z \in \partial B(2^{3n})$. The inequality (4.80) follows from (4.81) and (4.82).

Now, we need an upper bound for $\mathbb{P}[E_m \cap E_n]$, $m \leq n$. Clearly, $E_m \cap E_n \subset E'_m \cap E'_n$, and note that the event $E'_m \cap E'_n$ means that the particle hits $\partial B(2^{3n})$ before $\partial B(2^{3m-1})$ starting from a site on $\partial B(2^{3m})$, and then never hits $\partial B(2^{3n-1})$ starting from a site on $\partial B(2^{3n})$. So, again using (4.10) and Lemma 4.4, we write analogously to (4.82) (and also omitting a couple of lines of elementary calculations)

$$\begin{aligned}
\mathbb{P}[E_m \cap E_n] &\leq \mathbb{P}[E'_m \cap E'_n] \\
&= \frac{(a(2^{3m-1}))^{-1} - (a(2^{3m}))^{-1} + O(2^{-3m})}{(a(2^{3m-1}))^{-1} - (a(2^{3n}))^{-1} + O(2^{-3m})}
\end{aligned}$$

---

[13] E.g., by comparison/coupling to Brownian motion, or directly, in the spirit of Exercise 4.12.

$$\times \left(1 - \frac{a(2^{3n-1}) + O(2^{-3n})}{a(2^{3n}) + O(2^{-3n})}\right)$$

$$= \frac{1}{(3(n-m)+1)(3m+\gamma^*)}(1 + o(2^{-3m}))$$

$$\leq \frac{c_9}{m(n-m+1)}. \tag{4.83}$$

Therefore, we can write

$$\sum_{m,n=1}^{k} \mathbb{P}[E_i \cap E_j] \leq 2 \sum_{1 \leq m \leq n \leq k} \mathbb{P}[E_i \cap E_j]$$

$$\leq 2c_9 \sum_{1 \leq m \leq n \leq k} \frac{1}{m(n-m+1)}$$

$$= 2c_9 \sum_{m=1}^{k} \frac{1}{m} \sum_{n=m}^{k} \frac{1}{n-m+1}$$

$$\leq 2c_9 \sum_{m=1}^{k} \frac{1}{m} \sum_{n=m}^{k-m+1} \frac{1}{n-m+1}$$

$$= 2c_9 \left(\sum_{m=1}^{k} \frac{1}{m}\right)^2,$$

and so $\sum_{i,j=1}^{k} \mathbb{P}[E_i \cap E_j] \leq c_{10} \ln^2 k$. Now, (4.80) implies that $\sum_{i=1}^{k} \mathbb{P}[E_i] \geq c_{11} \ln k$, so, using (4.79), we obtain that

$$\mathbb{P}\left[\sum_{k=1}^{\infty} \mathbb{1}\{E_k\} = \infty\right] \geq c_{12} > 0.$$

To obtain that the probability in the preceding display must be equal to 1, we need a suitable 0-1 law. Conveniently enough, it is provided by proposition 3.8 in chapter 2 of [86]: if every set is either recurrent or transient with respect to $\widehat{S}$, then every tail event must have probability 0 or 1. Now, note that Theorem 4.11 implies that every set must be recurrent or transient indeed. This concludes the proof of Theorem 4.18. □

4.6 Exercises

Exercise 4.1. For one-dimensional random walk X_n with drift (i.e., it jumps to the left with probability $p \in (0, \frac{1}{2})$ and to the right with probability $1-p$), prove that (somewhat surprisingly) $|X_n|$ is a Markov chain, and calculate its transition probabilities.

Exercise 4.2. Calculate Green's function of the conditioned one-dimensional SRW.

Exercise 4.3. Consider a nearest-neighbour random walk on $\mathbb{Z}_+$ with drift towards the origin, $p(n, n + 1) = 1 - p(n, n - 1) = p < \frac{1}{2}$. What will be its conditioned (on never hitting the origin) version?

Exercise 4.4. Now consider a one-dimensional nearest-neighbour *Lamperti* random walk (recall Exercise 2.27): $p(n, n+1) = 1 - p(n, n-1) = \frac{1}{2} + \frac{c}{n}$ with $c < \frac{1}{4}$, so that the random walk is recurrent. What can you say about its conditioned version?

Exercise 4.5. Assume that the original Markov chain is reversible; must its h-transform be reversible as well?

Exercise 4.6. For any finite $A \subset \mathbb{Z}^2$, find a (nontrivial) nonnegative function which is zero on A and harmonic outside A.

Exercise 4.7. Prove that

$$\mathbb{P}_x[\hat{\tau}_N < \infty] = \frac{1}{a(x)} \qquad (4.84)$$

for any $x \in \mathbb{Z}^2 \setminus \{0\}$.

Exercise 4.8. Can you find an expression for Green's function of the h-transform in the general case (as in Section 4.1), i.e., an analogue of (4.12)? What else should we assume for that?

Exercise 4.9. Do we actually need recurrence in Lemma 4.9?

Exercise 4.10. Prove that the conditioned walk $\widehat{S}$ is uniformly elliptic, i.e., there exists $c > 0$ such that $\mathbb{P}_x[\widehat{S}_1 = y] > c$ for all $x, y \in \mathbb{Z}^2 \setminus \{0\}$ such that $x \sim y$.

Exercise 4.11. Can you find a heuristic explanation of (4.21) (i.e., that the probability that $\widehat{S}$ ever visits a very distant site is approximately $1/2$)?

Exercise 4.12. Give at least a couple of different proofs of (4.39) which do not use Lemma 3.11.

Exercise 4.13. Argue that it is evident that, for all $y \in \mathbb{Z}^2 \setminus \{0\}$,

$$\mathbb{E}_y(\widehat{G}(S_1, y) - \widehat{G}(y, y)) = 1. \qquad (4.85)$$

Nevertheless, verify (4.85) *directly* (i.e., using (4.12)).

Exercise 4.14. Prove that there exist two positive constants c_1, c_2 such that, for all $x, y \in \mathbb{Z}^2 \setminus \{0\}$,

$$\frac{c_1}{\ln(1 + \|x\| \vee \|y\|)} \leq \hat{g}(x, y) \leq \frac{c_2}{\ln(1 + \|x\| \vee \|y\|)}. \tag{4.86}$$

Exercise 4.15. With a "direct" computation (i.e., not using the general fact that Green's functions give rise to martingales), prove that the process $(\hat{\ell}(\widehat{S}_{n \wedge \hat{\tau}_y}, y), n \geq 0)$ (recall (4.17)) is a martingale.

Exercise 4.16. In Lemma 4.13, can one substitute the constant 5 by any fixed positive number?

Exercise 4.17. Let A be a finite subset of $\mathbb{Z}^2 \setminus \{0\}$. Prove that

$$\sum_{z \in A} \hat{g}(y_0, z) \widehat{\mathrm{hm}}_A(z) = \frac{1}{\widehat{\mathrm{cap}}(A)} \tag{4.87}$$

for any $y_0 \in A$.

Exercise 4.18. For any $\Lambda \subset (\mathbb{Z}^2 \setminus \{0\})$, and $x, y \in \Lambda$, quite analogously to (3.43) we can define

$$\widehat{G}_\Lambda(x, y) = \mathbb{E}_x \sum_{k=0}^{\hat{\tau}_{\Lambda^C} - 1} \mathbf{1}\{\widehat{S}_k = y\} \tag{4.88}$$

to be the mean number of visits of the conditioned walk to y starting from x before stepping out of Λ.

Prove that the analogue of (4.22) holds also for the restricted Green's function of the conditioned walk:

$$a^2(x)\widehat{G}_\Lambda(x, y) = a^2(y)\widehat{G}_\Lambda(y, x) \tag{4.89}$$

for all $x, y \in \Lambda$.

Exercise 4.19. Prove the analogue of Theorem 3.13: for any $\Lambda \subset (\mathbb{Z}^2 \setminus \{0\})$

$$\widehat{G}_\Lambda(x, y) = \widehat{G}(x, y) - \mathbb{E}_x \widehat{G}(\widehat{S}_{\hat{\tau}_{\Lambda^C}}, y) \tag{4.90}$$

for all $x, y \in \Lambda$.

Exercise 4.20. Check (4.65) and (4.66).

Exercise 4.21. Give examples of $A \subset \mathbb{Z}^2$ with $|A_n| \to \infty$ such that (recall that $\mathcal{R}(A)$ is the proportion of sites in A visited by the conditioned walk)

 (i) $\mathcal{R}(A_n)$ converges in distribution to Bernoulli$(1/2)$;
 (ii) $\mathcal{R}(A_n)$ converges in probability to $1/2$.

Exercise 4.22. Can we somehow obtain mixtures of the three main limit distributions of $\mathcal{R}(\cdot)$, and how shall we recognize their "domains of attraction"? For example, try to analyse the following case: let $A_n = \mathsf{B}(r_n e_1, n)$, where $r_n \to \infty$ *in some way.*

5

Intermezzo: soft local times and Poisson processes of objects

Here we (apparently) digress from our main topic of simple random walks, and discuss the following two subjects. One of them is the method of *soft local times*: it is an "algorithm" that permits us to obtain a realization of an adapted stochastic process on a general space Σ taking an auxiliary Poisson point process on $\Sigma \times \mathbb{R}_+$ as an input. It turns out that this method can be quite useful for dealing with sequences of random walk's excursions; as we remember from Section 4.5, analysing those is important for proving interesting results about our models. So, it is unsurprising that it will also come in handy later in Chapter 6.

Then in Section 5.2 we digress even more and discuss Poisson processes of *infinite* objects, using the Poisson line process as a basic example.[1] Given that the next chapter is in fact devoted to Poisson processes of specific infinite objects (which are random walk's trajectories), it looks like a good idea to spend some time on a simpler model of "similar" kind in order to build some intuition about what will come next.

5.1 Soft local times

Let us start with the following elementary question. Assume that X and Y are two random variables with the same support[2] but different distributions. Let $X_1, X_2, X_3, \ldots$ be a sequence of independent copies of X. Does there exist an infinite permutation[3] $\sigma = (\sigma(1), \sigma(2), \sigma(3), \ldots)$ such that the sequence $X_{\sigma(1)}, X_{\sigma(2)}, X_{\sigma(3)}, \ldots$ has the same law as the sequence $Y_1, Y_2, Y_3, \ldots$,

---

[1] In the author's experience, many people working with random walks do not know how Poisson line processes are constructed; so this example by itself might be interesting to some readers.

[2] Informally, the support of a random variable Z is the (minimal) set where it lives. Formally, it is the intersection of all *closed* sets F such that $\mathbb{P}[Z \in F] = 1$ (therefore, in particular, the support is a closed set).

[3] I.e., a bijection $\sigma \colon \mathbb{N} \mapsto \mathbb{N}$.

a sequence of independent copies of Y? Of course, such a permutation should be random: if it is deterministic, then the permuted sequence would simply have the original law.[4] For constructing σ, one is allowed to use additional random variables (independent of the X-sequence) besides the realization of the X-sequence itself. As far as the author knows, constructing the permutation without using additional randomness (i.e., when the permutation is a *deterministic* function of the *random* sequence $X_1, X_2, X_3, \ldots$) is still an open problem, a rather interesting one.

As usual, when faced with such a question, one tries a "simple" case first, to see if it gives any insight on the general problem. For example, take X to be Binomial$(n, \frac{1}{2})$ and Y to be discrete Uniform$[0, n]$. One may even consider the case when X and Y are Bernoullis, with different probabilities of success. How can one obtain σ in these cases?

After some thought, one will come with the following solution, simple and straightforward: just generate the i.i.d. sequence $Y_1, Y_2, Y_3, \ldots$ independently, then there is a permutation that sends X-sequence to the Y-sequence. Indeed (this argument works for any pair of discrete random variables with the same support), almost surely any possible value of X (and Y) occurs infinitely many times both in the X-sequence and the Y-sequence. It is then quite straightforward to see that there is a permutation that sends one sequence to the other.

Now, let us be honest with ourselves: this solution looks like cheating. In a way, it is simply *too easy*. Common wisdom tells us, however, that there ain't no such thing as a free solution; in this case, the problem is that the preceding construction does not work at all when the random variables are continuous. Indeed, if we generate the two sequences independently, then, almost surely, no element of the first sequence will be even present in the second one. So, a different approach is needed.

Later in this section, we will see how to solve such a problem using a sequence of i.i.d. exponential random variables as additional randomness. The solution will come out as an elementary application of the method of *soft local times*, the main subject of this section. Generally speaking, the method of soft local times is a way to construct an adapted stochastic process on a general space Σ, using an auxiliary Poisson point process on $\Sigma \times \mathbb{R}_+$.

Naturally, we assume that the reader knows what is a Poisson point process in $\mathbb{R}^d$ with (not necessarily constant) rate λ. If one needs to consider a

---

[4] Even more, the permutation σ should depend on $X_1, X_2, X_3, \ldots$; if it is independent of the X-sequence, it is still easy to check that the permuted sequence has the original law.

Poisson process on, say, $\mathbb{Z} \times \mathbb{R}$, then it is still easy to understand what exactly it should be (a union of Poisson processes on the straight lines indexed by the sites of $\mathbb{Z}$). In any case, all this fits into the Poissonian paradigm: what happens in a domain does not affect what is going on in a disjoint domain, the probability that there is exactly one point in a "small" domain of volume δ located "around" x is $\delta\lambda(x)$ (up to terms of smaller order), and the probability that there are at least two points in that small domain is $o(\delta)$. Here, the tradition dictates that the author cites a comprehensive book on the subject, so [85].

Coming back to the soft local times method, we mention that, in full generality, it was introduced in [79]; see also [25, 26] which contain short surveys of this method applied to constructions of excursion processes.[5] The idea of using projections of Poisson processes for constructions of other (point) processes is not new; see e.g. [49, 50]. The key tool of this method (Lemma 5.1) appears in [102] in a simpler form, and the motivating example we gave in the beginning of this section is also from that paper.

Next, we are going to present the key result that makes the soft local times possible. Over here, we call it "the magic lemma". Assume that we have a space Σ, which has enough structure[6] that permits us to construct a Poisson point process on Σ of rate μ, where μ is a measure on Σ.

Now, the main object we need is the Poisson point process on $\Sigma \times \mathbb{R}_+$, with rate $\mu \otimes dv$, where dv is the Lebesgue measure on $\mathbb{R}_+$. At this point, we have to write some formalities. In the next display, Ξ is a countable index set. We prefer not to use $\mathbb{Z}_+$ for the indexing, because we are not willing to fix any particular ordering of the points of the Poisson process for the reason that will become clear in a few lines. Let

$$\mathcal{M} = \Big\{ \eta = \sum_{\varrho \in \Xi} \delta_{(z_\varrho, v_\varrho)}; z_\varrho \in \Sigma, v_\varrho \in \mathbb{R}_+,$$

$$\text{and } \eta(K) < \infty \text{ for all compact } K \Big\}, \qquad (5.1)$$

be the set[7] of point configurations of this process. It is a general fact that one can canonically construct such a Poisson point process η; see e.g. proposition 3.6 on p. 130 of [85] for details of this construction.

The following result is our "magic lemma": it provides us with a way

---

[5] But this will be treated in a detailed way in Section 6.3.1.

[6] For example, the following is enough: let Σ be a locally compact and Polish metric space, and μ is a Radon measure (i.e., every compact set has finite μ-measure) on the measurable space $(\Sigma, \mathcal{B})$, where $\mathcal{B}$ is the Borel σ-algebra on Σ.

[7] Endowed with sigma-algebra $\mathcal{D}$ generated by the evaluation maps $\eta \mapsto \eta(A)$, $A \in \mathcal{B} \otimes \mathcal{B}(\mathbb{R})$.

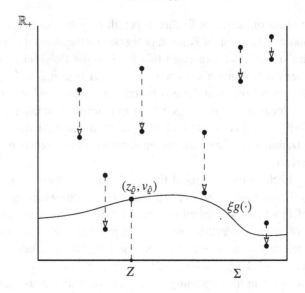

Figure 5.1 The magic lemma. The random variable Z has density g, •'s are points of η, and ∘'s are points of the new Poisson process η'.

to simulate a random element of Σ with law absolutely continuous with respect to μ, using the Poisson point process η. We first write it formally, and then explain what it means.

Lemma 5.1. *Let $g : \Sigma \to \mathbb{R}_+$ be a (nonnegative) measurable function such that $\int g(z)\mu(dz) = 1$. For $\eta = \sum_{\varrho \in \Xi} \delta_{(z_\varrho, v_\varrho)} \in M$, we define*

$$\xi = \inf\{t \geq 0; \ \text{there exists } \varrho \in \Xi \text{ such that } tg(z_\varrho) \geq v_\varrho\}. \qquad (5.2)$$

Then, under the law $\mathbb{Q}$ of the Poisson point process η,

1. *there exists a.s. a unique $\hat\varrho \in \Xi$ such that $\xi g(z_{\hat\varrho}) = v_{\hat\varrho}$,*
2. *$(z_{\hat\varrho}, \xi)$ is distributed as $g(z)\mu(dz) \otimes \mathrm{Exp}(1)$,*
3. *$\eta' := \sum_{\varrho \neq \hat\varrho} \delta_{(z_\varrho, v_\varrho - \xi g(z_\varrho))}$ has the same law as η and is independent of $(\xi, \hat\varrho)$.*

That is, in plain words (see Figure 5.1):

- In (5.2) we define ξ as the smallest positive number such that there is exactly one point $(z_{\hat\varrho}, v_{\hat\varrho})$ of the Poisson process on the graph of $\xi g(\cdot)$, and nothing below this graph.

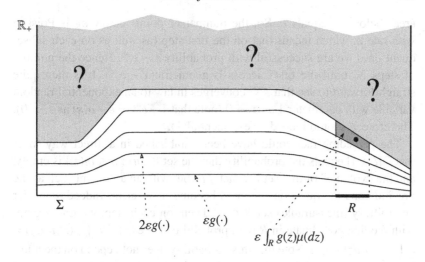

Figure 5.2 Slow exploration of the space: why Lemma 5.1 is valid.

- The first coordinate Z of the chosen point is a random variable with density g (with respect to μ). Also, ξ is exponential with parameter 1, and it is independent of Z.
- Remove the point that was chosen, and translate every other point (z, v) of η down by amount $\xi g(z)$. Call this new configuration η'. Then η' is also a Poisson point process on $\Sigma \times \mathbb{R}_+$ with rate $\mu \otimes dv$, and it is independent of ξ and Z.

Sketch of the proof of Lemma 5.1. The formal proof can be found in [79] (Lemma 5.1 is proposition 4.1 of [79]), and here we give only an informal argument to convince the reader that Lemma 5.1 is not only magic, but also true. In fact, this result is one of those statements that become evident after one thinks about it for a couple of minutes; so, it may be a good idea for the reader to ponder on it for some time before going further.

So, one may convince oneself that the result is valid e.g. in the following way. Fix a very small $\varepsilon > 0$ and let us *explore* the space as shown in Figure 5.2. That is, first look at the domain $\{(z, u) : u \leq \varepsilon g(z)\}$ and see if we find a point of the Poisson process there (observe that finding *two* points is highly improbable). If we don't, then we look at the domain $\{(z, u) : \varepsilon g(z) < u \leq 2\varepsilon g(z)\}$, and so on.

How many steps do we need to discover the first point? First, observe that g is a density, so it integrates to 1 with respect to μ, and therefore the area[8] below εg equals ε. So, the number of points below εg is Poisson with rate ε, which means that on the first step (as well as on each subsequent one) we are successful with probability $1 - e^{-\varepsilon}$. Hence the number of steps N_ε until the first success is geometric$(1 - e^{-\varepsilon})$. It is then quite straightforward to see that $\varepsilon N_\varepsilon$ converges in law to an exponential random variable with parameter 1 as $\varepsilon \to 0$ (note that $1 - e^{-\varepsilon} = \varepsilon + o(\varepsilon)$ as $\varepsilon \to 0$). Therefore, ξ should indeed be exponential(1).

The preceding fact could have been established in a direct way (note that $\mathbb{Q}[\xi > t]$ equals the probability that the set $\{(z, u) : u \le tg(z)\}$ is empty, and the "volume" of that set is exactly t), but with an argument such as the previous one the questions about Z become more clear. Indeed, consider an arbitrary (measurable) set $R \subset \Sigma$. Then, on each step, we find a point with the first coordinate in R with probability $1 - \exp\left(- \varepsilon \int_R g(z)\mu(dz)\right) = \varepsilon \int_R g(z)\mu(dz) + o(\varepsilon)$. Note that this probability does not depend on the number of steps already taken; that is, independently of the past, the conditional probability of finding a point with first coordinate in R given that something is found on the current step[9] is roughly $\int_R g(z)\mu(dz)$. This shows that ξ and Z are independent random variables.

As for the third part, simply observe that, at the time we discovered the first point, the shaded part on Figure 5.2 is still completely unexplored, and so its contents are independent of the pair (ξ, Z). In other words, we have a Poisson process on the set $\{(z, v) : v > \xi g(z)\}$ with the same rate, which can be transformed to the Poisson process in $\Sigma \times \mathbb{R}_+$ by subtracting $\xi g(\cdot)$ from the second coordinate (observe that such transformation is volume preserving).   □

Now, the key observation is that Lemma 5.1 allows us to construct virtually *any* discrete-time stochastic process! Moreover, one can effectively couple two or more stochastic processes using the same realization of the Poisson process. One can better visualize the picture in a continuous space, so, to give a clear idea of how the method works, assume that we desire to obtain a realization of a sequence of (not necessarily independent nor Markovian) random variables $X_1, X_2, X_3, \ldots$ taking values in the interval $[0, 1]$. Let us also construct *simultaneously* the sequence $Y_1, Y_2, Y_3, \ldots$, where (Y_k) are i.i.d. Uniform$[0, 1]$ random variables, thus effectively obtaining a *coupling* of the X- and Y-sequences. We assume that the law of X_k conditioned on $\mathcal{F}_{k-1}$ is a.s. absolutely continuous with respect to the

---

[8] With respect to $\mu \otimes dv$.

[9] That is, we effectively condition on ξ, and show that the conditional law of Z does not depend on the value of ξ.

Lebesgue measure on $[0, 1]$, where $\mathcal{F}_{k-1}$ is the sigma-algebra generated by $X_1, \ldots, X_{k-1}$.

This idea of using the method of soft local times to couple general stochastic processes with independent sequences has already proved to be useful in many situations; for this book, it will be useful as well, as we will see in Chapter 6.

Our method for constructing a coupling of the X- and Y-sequences is illustrated in Figure 5.3. Consider a Poisson point process in $[0, 1] \times \mathbb{R}_+$ with rate 1. Then one can obtain a realization of the Y-sequence by simply ordering the points according to their second coordinates, and then taking $Y_1, Y_2, Y_3, \ldots$ to be the first coordinates of these points. Now, to obtain a realization of the X-sequence using the same Poisson point process, one proceeds as follows.

- First, take the density $g(\cdot)$ of X_1 and multiply it by the unique positive number ξ_1 so that there is exactly one point of the Poisson process lying on the graph of $\xi_1 g$ and nothing strictly below it; X_1 is then the first coordinate of that point.
- Using Lemma 5.1, we see that, if we remove the point chosen on the previous step[10] and then translate every other point (z, u) of the Poisson process to $(z, u - \xi_1 g(z))$, then we obtain a Poisson process in $[0, 1] \times \mathbb{R}_+$ which is independent of the pair (ξ_1, X_1).
- Thus, we are ready to use Lemma 5.1 again in order to construct X_2.
- So, consider the conditional density $g(\cdot \mid \mathcal{F}_1)$ of X_2 given $\mathcal{F}_1$ and find the smallest positive number ξ_2 in such a way that exactly one point lies on the graph of $\xi_2 g(\cdot \mid \mathcal{F}_1) + \xi_1 g(\cdot)$ and exactly one (the point we picked first) below it; again, X_2 is the first coordinate of the point that lies on the graph.
- Continue with $g(\cdot \mid \mathcal{F}_2)$, and so on.

The fact that the X-sequence obtained in this way has the prescribed law is readily justified by the subsequent application of Lemma 5.1. Now let us state the formal result (it corresponds to proposition 4.3 of [79]); here it is only a bit more general since we formulate it for adapted processes. For a stochastic process $(Z_n, n \geq 0)$ adapted to a filtration $(\mathcal{F}_n, n \geq 0)$ we define

$$\xi_1 = \inf \{t \geq 0 : \text{ there exists } \varrho \in \Xi \text{ such that } tg(z_\varrho) \geq v_\varrho\},$$

$$\mathcal{G}_1(z) = \xi_1 g(z \mid \mathcal{F}_0), \text{ for } z \in \Sigma,$$

---

[10] This point has coordinates $(X_1, \xi_1 g(X_1))$.

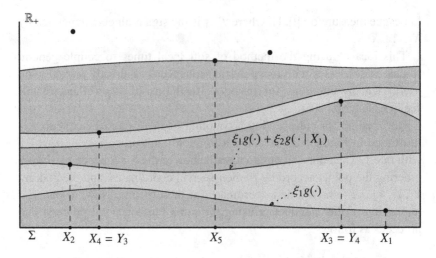

Figure 5.3 Soft local times: the simultaneous construction of the processes X and Y (here, $X_k = Y_k$ for $k = 1, 2, 5$); it is very important to observe that the points of the two processes need not necessarily appear in the same order with respect to the vertical axis.

where $g(\cdot \mid \mathcal{F}_0)$ is the density of Z_1 given $\mathcal{F}_0$, and

$$(z_1, v_1) \text{ is the unique pair in } \{(z_\varrho, v_\varrho)\}_{\varrho \in \Xi} \text{ with } \mathcal{G}_1(z_1) = v_1. \tag{5.3}$$

Denote also $R_1 = \{(z_1, v_1)\}$. Then for $n \geq 2$ we proceed inductively,

$$\xi_n = \inf \{t \geq 0 : \text{there exists } (z_\varrho, v_\varrho) \notin R_{n-1}$$
$$\text{such that } \mathcal{G}_{n-1}(z_\varrho) + tg(z_\varrho \mid \mathcal{F}_{n-1}) \geq v_\varrho\}, \tag{5.4}$$
$$\mathcal{G}_n(z) = \mathcal{G}_{n-1}(z) + \xi_n g(z \mid \mathcal{F}_{n-1}),$$

and

$$(z_n, v_n) \text{ is the unique pair } (z_\varrho, v_\varrho) \notin R_{n-1} \text{ with } \mathcal{G}_n(z_\varrho) = v_\varrho; \tag{5.5}$$

also, set $R_n = R_{n-1} \cup \{(z_n, v_n)\}$. These random functions $(\mathcal{G}_n, n \geq 1)$ deserve a special name:

Definition 5.2. We call $\mathcal{G}_n$ the *soft local time* of the process, at time n, with respect to the reference measure μ.

The previous discussion implies that the following result holds:

Proposition 5.3. *It holds that*

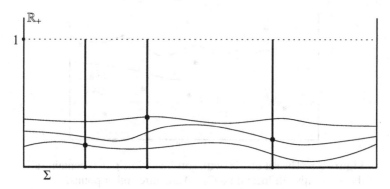

Figure 5.4 Comparison of soft and usual local times: the usual local time $L_3(\cdot)$ has three peaks (of size 1) at points z_1, z_2, z_3, and equals 0 in all other points. The soft one looks much softer.

(i) $(z_1, \ldots, z_n) \stackrel{law}{=} (Z_1, \ldots, Z_n)$ *and they are independent from* $\xi_1, \ldots, \xi_n$,

(ii) the point process

$$\sum_{(z_\varrho, v_\varrho) \notin R_n} \delta_{(z_\varrho, v_\varrho - \mathcal{G}_n(z_\varrho))}$$

is distributed as η *and independent of* R_n *and* $\xi_1, \ldots, \xi_n$,

for all $n \geq 1$.

To justify the choice of the name "soft local time" for the previously defined object, consider a stochastic process in a finite or countable state space, and define the "usual" local time of the process by

$$L_n(z) = \sum_{k=1}^{n} \mathbf{1}\{X_k = z\}. \tag{5.6}$$

Now, just look at Figure 5.4.

Next, we establish a very important relation between these two different local times: their expectations are equal.

Proposition 5.4. *For all* $z \in \Sigma$, *it holds that*

$$\mathbb{E}\mathcal{G}_n(z) = \mathbb{E}L_n(z) = \sum_{k=1}^{n} \mathbb{P}[X_k = z]. \tag{5.7}$$

Notice that in continuous space we cannot expect the preceding result to be true, since typically $\mathbb{E}L_n(z)$ would be just 0 for any z. Nevertheless, an analogous result holds in the general setting as well (cf. theorem 4.6

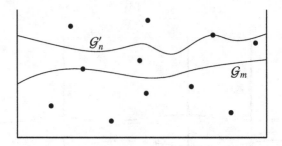

Figure 5.5 The set of Y's (with soft local time $\mathcal{G}'_n(\cdot)$) contains all the X's (with soft local time $\mathcal{G}_m(\cdot)$) and three other points.

of [79]), but, to formulate it properly, one would need to define the so-called *expected local time density* first (cf. (4.16) of [79]), which we prefer not to do here.

Proof of Proposition 5.4. It is an easy calculation that uses conditioning and induction. First, observe that $g(z \mid \mathcal{F}_{n-1}) = \mathbb{P}[X_n = z \mid \mathcal{F}_{n-1}]$, so we have

$$\mathbb{E}\mathcal{G}_1(z) = \mathbb{E}(\mathbb{P}[X_1 = z \mid \mathcal{F}_0]) = \mathbb{P}[X_1 = z] = \mathbb{E}L_1(z).$$

Then, we proceed by induction: note that $\mathcal{G}_{n-1}(z)$ is $\mathcal{F}_{n-1}$-measurable, and ξ_n is a mean-1 random variable which is independent of $\mathcal{F}_{n-1}$. Recall also (5.4) and write

$$\begin{aligned}
\mathbb{E}\mathcal{G}_n(z) &= \mathbb{E}(\mathbb{E}(\mathcal{G}_n(z) \mid \mathcal{F}_{n-1})) \\
&= \mathbb{E}\mathcal{G}_{n-1}(z) + \mathbb{E}(\mathbb{E}(\xi_n g(z \mid \mathcal{F}_{n-1}) \mid \mathcal{F}_{n-1})) \\
&= \mathbb{E}\mathcal{G}_{n-1}(z) + \mathbb{E}(g(z \mid \mathcal{F}_{n-1})\mathbb{E}(\xi_n \mid \mathcal{F}_{n-1})) \\
&= \mathbb{E}\mathcal{G}_{n-1}(z) + \mathbb{E}(\mathbb{P}[X_n = z \mid \mathcal{F}_{n-1})) \\
&= \mathbb{E}\mathcal{G}_{n-1}(z) + \mathbb{P}[X_n = z].
\end{aligned}$$

This concludes the proof.   □

As mentioned before, soft local times work really well for couplings of stochastic processes: indeed, simply construct them in the way described earlier using *the same* realization of the Poisson point process. Observe that for this coupling of the processes (X_n) and (Y_n), it holds that

$$\mathbb{P}[\{X_1, \ldots, X_m\} \subset \{Y_1, \ldots, Y_n\}] \geq \mathbb{P}[\mathcal{G}_m(z) \leq \mathcal{G}'_n(z) \text{ for all } z \in \Sigma], \quad (5.8)$$

where $\mathcal{G}'$ is the soft local time of Y; see Figure 5.5. Then, in principle, one may use large deviations tools to estimate the right-hand side of (5.8). One has to pay attention to the following, though: it is easy to see that the random variables $(\xi_1, \ldots, \xi_m)$ are not independent of $(\xi'_1, \ldots, \xi'_n)$ (which enter to G'_n). This can be usually circumvented in the following way: we find a *deterministic* function $\varphi \colon \Sigma \mapsto \mathbb{R}$, which should typically be "between" G_m and G'_n, and then write

$$\mathbb{P}[G_m(z) \leq G'_n(z) \text{ for all } z \in \Sigma]$$

$$\geq \mathbb{P}[G_m(z) \leq \varphi(z) \text{ for all } z \in \Sigma] + \mathbb{P}[\varphi(z) \leq G'_n(z) \text{ for all } z \in \Sigma] - 1. \quad (5.9)$$

Note that in the right-hand side of the preceding relation, we do not have this "conflict of ξ's" anymore. Let us also mention that in the previous large deviation estimates, one has to deal with sequences of *random functions* (not just real-valued random variables). When the state space Σ is finite, this difficulty can be usually circumvented by considering the values of the functions separately in each point of Σ and then using the union bound, hoping that this last step will not cost too much. Otherwise, one has to do the large deviations for random functions directly using some advanced tools from the theory of empirical processes[11]; see e.g. section 6 of [30] and lemma 2.9 of [23] for examples of how large deviations for soft local times may be treated.

To underline the importance of finding couplings as before (and also the *exact* couplings of Section 5.1.1), observe that there are many quantities of interest that can be expressed in terms of local times (as defined in (5.6)) only (that is, they do not depend on the order). Such quantities are, for example,

- hitting time of a site x: $\tau(x) = \min\{n : L_n(x) > 0\}$;
- cover time: $\min\{n : L_n(x) > 0 \text{ for all } x \in \Sigma\}$, where Σ is the space where the process lives;
- blanket time [36]: $\min\{n \geq 1 : L_n(x) \geq \delta n\pi(x)\}$, where π is the stationary measure of the process and $\delta \in (0, 1)$ is a parameter;
- disconnection time [32, 93]: loosely speaking, it is the time n when the set $\{x : L_n(x) > 0\}$ becomes "big enough" to "disconnect" the space Σ in some precise sense;
- the set of favorite (most visited) sites (e.g. [52, 101]): $\{x : L_n(x) \geq L_n(y) \text{ for all } y \in \Sigma\}$;
- and so on.

Now, finally, let us go back to the example from the beginning of this

---

[11] Note that they have to be more advanced than Bousquet's inequality (see theorem 12.5 of [12]) since, due to these i.i.d. exponential ξ's, the terms are not a.s. bounded.

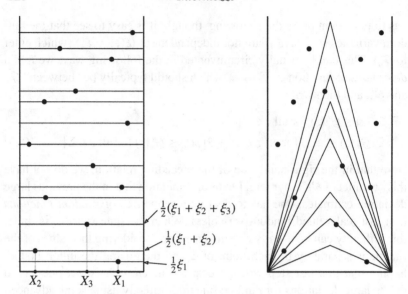

Figure 5.6 Making uniforms triangular. We first obtain a
particular instance of the Poisson process in $[-1, 1] \times \mathbb{R}_+$ using
the X-sequence, and then use the same collection of points to
build the Y-sequence. It holds that $\sigma(1) = 1$, $\sigma(2) = 3$, $\sigma(3) = 2$,
$\sigma(4) = 6$, $\sigma(5) = 4$, $\sigma(6) = 10$, $\sigma(7) = 5$ (i.e., $Y_7 = X_5$, etc.).

section: recall that we had a realization of an i.i.d. sequence $X_1, X_2, X_3, \dots$,
and wanted to find an infinite permutation σ such that $X_{\sigma(1)}, X_{\sigma(2)}, X_{\sigma(3)}, \dots$
is also an i.i.d. sequence, however, sampled from another distribution (with
the same support). With Proposition 5.3 at hand, the solution is relatively
simple. Let $\xi_1, \xi_2, \xi_3, \dots$ be a sequence of i.i.d. exponential(1) random variables; this sequence will serve as an additional randomness. As an example, let us consider the case when X is uniform on $[-1, 1]$, and Y has the
"triangular" density $f(y) = (1 - |y|)\mathbf{1}\{y \in [-1, 1]\}$. The first step is to *re-
construct* a Poisson process in $[-1, 1] \times \mathbb{R}_+$, using X's and ξ's. This can
be done in the following way (see Figure 5.6): for all $n \geq 1$, put a point
to $(X_n, \frac{1}{2}(\xi_1 + \cdots + \xi_n))$. Then, using this Poisson process, we obtain the
sequence $Y_1, Y_2, Y_3, \dots$ of i.i.d. triangular random variables in the way described previously; look at Figure 5.6, which speaks for itself. Clearly, one
sequence is a permutation of the other: they use the same points! We leave
as an exercise for the reader to check that, *this time*, essentially the same
solution works in the general case.

5.1.1 Exact matchings with soft local times

So far, we have seen how the method of soft local times is useful for *dominating* the range of one process with the range of another one (recall (5.9)). This is indeed the main application of this method that one can find in the literature. Still, a natural question that one may ask is: assume that the processes X and Y are "close" in some sense. Is it possible to *match* their ranges exactly *at finite time* (i.e., not as in the beginning of this chapter), i.e., is it possible to couple these processes in such a way that $(X_1, \ldots, X_n)$ and $(Y_1, \ldots, Y_n)$ are just permutations of each other with high probability[12]?

The problem, however, is that the "vanilla" soft local time method will not work for finding exact matchings of this sort. The reason is that, while the soft local times $\mathcal{G}_n^X$ and $\mathcal{G}_n^Y$ are *relatively* close, with high probability they will still be locally different, so that there will be points of the Poisson process which are not included to the range of X but included to the range of Y, and vise versa. However, it turns out to be possible to modify this method to make it work for this purpose. In the following, we explain the ideas of [30, 31]. We will formulate the results in greater generality, since we do not intend to prove them here anyway.

Definition 5.5. Let (Σ, ρ) be a compact metric space, with $\mathcal{B}(\Sigma)$ representing its Borel σ-algebra. We say that (Σ, ρ) is of *polynomial class*, when there exist some $\beta \geq 0$ and $\varphi \geq 1$ such that, for all $r \in (0, 1]$, the number of open balls of radius at most r needed to cover Σ is smaller than or equal to $\varphi r^{-\beta}$.

As an example of metric space of polynomial class, consider first a finite space Σ, endowed with the discrete metric

$$\rho(x, y) = \mathbf{1}\{x \neq y\}, \text{ for } x, y \in \Sigma.$$

In this case, we can choose $\beta = 0$ and $\varphi = |\Sigma|$. As a second example, let Σ be a compact k-dimensional Lipschitz submanifold of $\mathbb{R}^m$ with metric induced by the Euclidean norm of $\mathbb{R}^m$. In this case, we can take $\beta = k$, but φ will in general depend on the precise structure of Σ. It is important to observe that, for a finite Σ, it may not be the best idea to use the preceding discrete metric; one may be better off with another one, e.g., the metric inherited from the Euclidean space where Σ is immersed (see e.g. the proof of lemma 2.9 of [23]).

Here, we consider a Markov chain $X = (X_i)_{i \geq 1}$ on a *general* (not necessarily countable) state space Σ. Such a process is characterized by its

---

[12] As a motivating example, see Exercise 5.5.

transition kernel $\mathfrak{P}(x, dy)$ and initial distribution $\mathfrak{v}$, on $(\Sigma, \mathcal{B}(\Sigma))$: for any $B \in \mathcal{B}(\Sigma)$ and $n \geq 0$,

$$\mathbb{P}[X_{n+1} \in B \mid X_n = x, X_{n-1}, \ldots, X_0] = \mathbb{P}[X_{n+1} \in B \mid X_n = x]$$
$$= \mathfrak{P}(x, B),$$

and

$$\mathbb{P}[X_0 \in B] = \mathfrak{v}(B).$$

Let us suppose that the chain has a unique invariant probability measure Π, that is,

$$\int_\Sigma \mathfrak{P}(x, B) \, \Pi(dx) = \Pi(B)$$

for all $B \in \mathcal{B}(\Sigma)$. Moreover, we assume that the initial distribution and the transition kernel are absolutely continuous with respect to Π. Let us denote respectively by $v(\cdot)$ and $\tilde{p}(x, \cdot)$ the Radon–Nikodym derivatives (i.e., *densities*) of $\mathfrak{v}(\cdot)$ and $\mathfrak{P}(x, \cdot)$: for all $A \in \mathcal{B}(\Sigma)$

$$\mathfrak{v}(A) = \int_A v(y) \, \Pi(dy),$$
$$\mathfrak{P}(x, A) = \int_A \tilde{p}(x, y) \, \Pi(dy), \text{ for } x \in \Sigma.$$

Let us also use assume that the density $\tilde{p}(x, \cdot)$ is *uniformly Hölder continuous*, that is, there exist constants $\kappa > 0$ and $\gamma \in (0, 1]$ such that for all $x, z, z' \in \Sigma$,

$$|\tilde{p}(x, z) - \tilde{p}(x, z')| \leq \kappa(\rho(z, z'))^\gamma. \tag{5.10}$$

Next, let us suppose that there exists $\varepsilon \in (0, \frac{1}{2}]$ such that

$$\sup_{x,y \in \Sigma} |\tilde{p}(x, y) - 1| \leq \varepsilon, \tag{5.11}$$

and

$$\sup_{x \in \Sigma} |v(x) - 1| \leq \varepsilon. \tag{5.12}$$

Observe that (5.12) is not very restrictive because, due to (5.11), the chain will anyway come quite close to stationarity already on step 2.

Additionally, let us denote by $Y = (Y_i)_{i \geq 1}$ a sequence of i.i.d. random variables with law Π.

Before stating the main result of this subsection, we recall the definition of the total variation distance between two probability measures $\bar{\mu}$ and $\hat{\mu}$ on some measurable space $(\Omega, \mathcal{T})$:

$$d_{TV}(\bar{\mu}, \hat{\mu}) = \sup_{A \in \mathcal{T}} |\bar{\mu}(A) - \hat{\mu}(A)|.$$

When dealing with random elements U and V, we will write (with a slight abuse of notation) $d_{TV}(U, V)$ to denote the total variation distance between the laws of U and V. Denoting by $L_n^Z := (L_n^Z(x))_{x \in \Sigma}$ the local time field of the process $Z = X$ or Y at time n, we are now ready to state the following:

Theorem 5.6. *Suppose that (Σ, ρ) is of polynomial class with parameters β, φ, that the transition density is uniformly Hölder continuous in the sense of (5.10), and also that both the transition density and the initial distribution are close to stationarity in the sense of (5.11)–(5.12). Then there exists a universal positive constant K such that, for all $n \geq 1$, it holds that*

$$d_{TV}(L_n^X, L_n^Y) \leq K\varepsilon \sqrt{1 + \ln(\varphi 2^\beta) + \frac{\beta}{\gamma} \ln\left(\frac{\kappa \vee (2\varepsilon)}{\varepsilon}\right)}. \tag{5.13}$$

Notice that, in fact, the estimate (5.13) is only useful for small enough ε (recall that the total variation distance never exceeds 1). Although not every Markov chain comes close to the stationary distribution in just one step, that can be sometimes circumvented by considering the process at times $k, 2k, 3k, \ldots$ with a large k. Nevertheless, it is also relevant to check if we can obtain a uniform control in n of $d_{TV}(L_n^X, L_n^Y)$ away from the "almost stationarity" regime. We state the following:

Theorem 5.7. *As before, assume that (Σ, ρ) is of polynomial class and that the transition density is uniformly Hölder continuous. Assume also that*

$$\theta := \left(\sup_{x,y \in \Sigma} |\tilde{p}(x, y) - 1| \right) \vee \left(\sup_{x \in \Sigma} |\nu(x) - 1| \right) < 1. \tag{5.14}$$

Then there exists $K' = K'(\beta, \varphi, \kappa, \gamma, \theta) > 0$, decreasing in θ, such that

$$d_{TV}(L_n^X, L_n^Y) \leq 1 - K'$$

for all $n \geq 1$.

Such a result can be useful e.g. in the following context: if we are able to prove that, for the i.i.d. sequence, something happens with probability close to 1, then the same happens for the field of local times of the Markov chain with at least uniformly positive probability. Observe that it is not unusual that the fact that the probability of something is uniformly positive

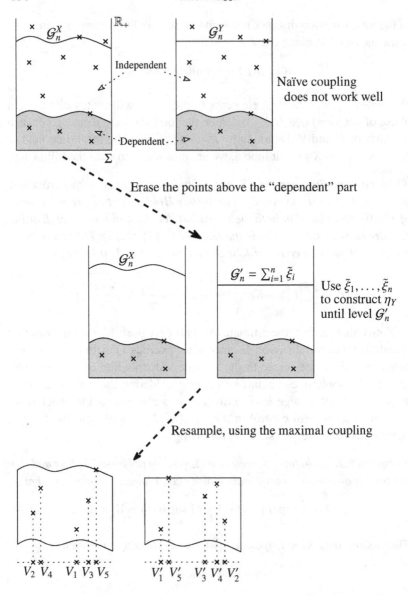

Figure 5.7 Resampling of the "independent parts".

implies that it should be close to 1 then (because one frequently has general results stating that this something should converge to 0 or 1).

Now we briefly describe the main idea of the proof of the preceding results (and suggest the interested reader to look at [30, 31] for a complete argument); this idea is also depicted in Figure 5.7 (we refer to [30] for some notations used there; in that paper, G was used to denote soft local times, but here this notation is reserved for Green's function).

- As noted in the beginning of this subsection, the "direct" soft local time coupling will likely not work (look at the two crosses in the top-right corners of the first part of the picture).
- Since the Markov chain X is already close to the equilibrium, it will "regenerate" quite often, and it is therefore possible to extract an "independent part" from $(X_1, \ldots, X_n)$ (the elements of this independent part will be i.i.d. with law Π).
- Then, to generate the local times up to time n, we may do it first for the "dependent part", and then for the "independent" one.
- Note that, to obtain the i.i.d. sequence $(Y_1, \ldots, Y_n)$, their soft local time must be "flat", as pictured on the right part of Figure 5.7. Now, here comes the key idea: let us erase the points above the soft local time of the dependent part, and then try to "resample" the two Poisson processes in such a way that the *projections* (although not necessarily the points themselves) coincide with high probability. This will amount to constructing couplings of *binomial processes* (local times of i.i.d. random elements on some common space) with *slightly* different laws; proposition 5.1 of [30] shows that this is indeed possible. If that coupling of binomial processes is successful, then the local times of $(X_1, \ldots, X_n)$ and $(Y_1, \ldots, Y_n)$ will coincide.

The author understands that the preceding does not even remotely qualify to be a proof; he has a positive experience, though, of showing this argument to some colleagues and hear, "ah yes, sure!" as a response.

5.2 Poisson processes of objects

Of course, all people know what a Poisson process of points in $\mathbb{R}^d$ is. But what if we need a Poisson process of more complicated objects, which still live in $\mathbb{R}^d$? What is the *right* way to define it? Naturally, we need the picture to be invariant with respect to isometries.[13] Also, it should be, well, *as independent as possible*, whatever it may mean.

Observe that, if those objects are bounded (not necessarily uniformly), one can use the following natural procedure: take a d-dimensional Poisson

---

[13] Those are translations, rotations, reflections, and combinations of them.

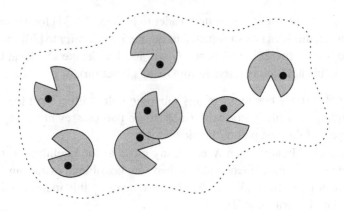

Figure 5.8 A Poisson process of *finite* objects.

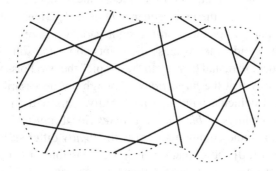

Figure 5.9 A Poisson line process (observed in the dotted domain) in $\mathbb{R}^2$.

point process of rate $\lambda > 0$, and "attach" the objects to the points independently (as e.g. in Figure 5.8). A broad example of this is the Poisson Boolean model, cf. e.g. [67]; there, these objects are balls/disks with random radii.

However, the situation becomes much more complicated if we need to build a Poisson process of *infinite* objects. For example, what about a two-dimensional Poisson process of lines, which should look like the example shown in Figure 5.9?

An idea that first comes to mind is simply to take a two-dimensional Poisson point process, and draw independent lines in random uniform directions through each point. One quickly realizes, however, that this way we would rather see what is shown in Figure 5.10: there will be *too many* lines, one would obtain a dense set on the plane instead of the nice picture

Figure 5.10 Too many lines!

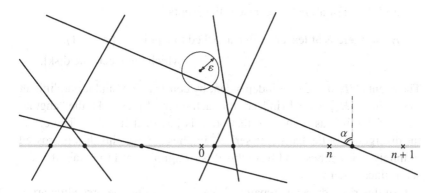

Figure 5.11 Constructing a Poisson line process using the reference line (here, $\alpha \in [-\pi/2, \pi/2]$ is the angle between the line and the normal vector).

shown in Figure 5.9. Another idea can be the following: first, fix a straight line on the plane (it can be the horizontal axis or just anything; it is the thicker line in Figure 5.11), and then consider a one-dimensional Poisson point process on this line. Then, through each of these points, draw a line with uniformly distributed direction (that is, α in Figure 5.11 is uniform in $[-\pi/2, \pi/2]$; for definiteness, think that the positive values of α are on the left side with respect to the normal vector pointing up) independently, thus obtaining the "process of lines" (not including the "reference" line) in $\mathbb{R}^2$.

The preceding looks as a reasonable procedure, but, in fact, it is not. Let us show that, as a result, we obtain a dense set again. Assume without loss of generality that the reference line is the horizontal axis, and consider a disk of radius $\varepsilon > 0$ situated somewhere above the origin (as in

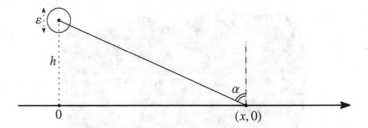

Figure 5.12 Calculating the expected number of lines that intersect the small ball. It holds that $\alpha = \arccos \frac{h}{\sqrt{x^2+h^2}}$.

Figure 5.11). For all $n \in \mathbb{Z}$, consider the events

$$H_n = \{\text{there is at least one line attached to a point in } [n, n+1),$$

$$\text{which intersects the disk}\}.$$

The events $(H_n, n \in \mathbb{Z})$ are independent by construction, and it not difficult to see that $\mathbb{P}[H_n] \asymp \varepsilon n^{-1}$ (indeed, for each point of $[n, n+1)$, the "angular size" of the disk as seen from that point is just of that order). Therefore, the divergence of the harmonic series[14] implies that a.s. this disk is crossed infinitely many times, and from this it is straightforward to obtain that the set of lines is dense.

Can this procedure be "repaired"? Well, examining the preceding argument, we see that the problem was that we gave "too much weight" to the angles which are close to $\pm\pi/2$. Therefore, choosing the direction uniformly does not work, and hence we need to choose it with some other density $\varphi(\cdot)$ on $[-\pi/2, \pi/2]$ (of course, it should be symmetric with respect to 0, i.e., the direction of the normal).

What should be this φ? Consider a small disk of diameter ε situated at distance h above the origin, as in Figure 5.12. Consider a point $(x, 0)$ on the reference line (horizontal axis), with $x > 0$. Then, clearly, to intersect the disk, the direction of a straight line passing through x must be in $[\alpha - \frac{\delta}{2}, \alpha + \frac{\delta}{2}]$, where $\alpha = \arccos \frac{h}{\sqrt{x^2+h^2}}$ and (up to terms of smaller order) $\delta = \frac{\varepsilon}{\sqrt{x^2+h^2}}$. So, if λ is the rate of the Poisson point process on the reference line and $N(h, \varepsilon)$ is the mean number of lines intersecting the small ball, we have

$$\mathbb{E}N(h, \varepsilon) = \lambda\varepsilon \int_{-\infty}^{+\infty} \frac{\varphi(\arccos \frac{h}{\sqrt{x^2+h^2}})}{\sqrt{x^2 + h^2}} \, dx + o(\varepsilon). \tag{5.15}$$

---

[14] This again! We meet the harmonic series quite frequently in two dimensions…

This does not look very nice, but notice that, if we just erase "φ" and "arccos" from (5.15),[15] the integral would become something more familiar (recall the Cauchy density)

$$\int\limits_{-\infty}^{+\infty} \frac{h}{x^2 + h^2}\, dx = \int\limits_{-\infty}^{+\infty} \frac{1}{(\frac{x}{h})^2 + 1}\, d\Big(\frac{x}{h}\Big) = \int\limits_{-\infty}^{+\infty} \frac{du}{u^2 + 1} = \pi,$$

so the parameter h disappears. And, actually, it is easy to get rid of φ and arccos at once: just choose $\varphi(\alpha) = \frac{1}{2}\cos\alpha$. So, we obtain from (5.15) that $\mathbb{E}N(h,\varepsilon) = \frac{1}{2}\pi\lambda\varepsilon + o(\varepsilon)$, which is a good sign that $\varphi(\alpha) = \frac{1}{2}\cos\alpha$ may indeed work for defining the Poisson line process.

The preceding construction is obviously invariant with respect to translations *in the direction of the reference line*, and, *apparently*, in the other directions too (there is no dependence on h for the expectations, but still some formalities are missing), but what about the rotational invariance? This can be proved directly,[16] but, instead of doing this now, let us consider another (more general[17]) approach to defining Poisson processes of objects. The idea is to represent these objects as *points in the parameter space*; i.e., each possible object is described by a (unique) set of parameters, chosen in some convenient (and clever!) way. Then we just take a Poisson *point* process in that parameter space, which is a process of objects naturally.

So, how can one carry this out in our case? Remember that we already constructed something translationally invariant, so let us try to find a parameter space where the rotational invariance would naturally appear. Note that any straight line that does not pass through the origin can be uniquely determined by two parameters: the distance r from the line to the origin, and the angle θ between the horizontal axis and the shortest segment linking the line to the origin. So, the idea is to take a realization of a Poisson point process (with some constant rate) in the parameter space $\mathbb{R}_+ \times [0, 2\pi)$, and translate it to a set of lines in $\mathbb{R}^2$, as shown in Figure 5.13.

Now, what kind of process do we obtain? First, it is clearly invariant under rotations. Secondly, it is not so obvious that it should be invariant with respect to translations. Instead of trying to prove it directly, we prefer to show that this construction is equivalent to the one with reference line (and hence get the translational invariance for free). Indeed, assume again that the reference line is the horizontal axis. Then (look at Figure 5.14) we

---

[15] And the parentheses as well, although it not strictly necessary.

[16] Please, try to do it!

[17] In fact, it is *the* approach.

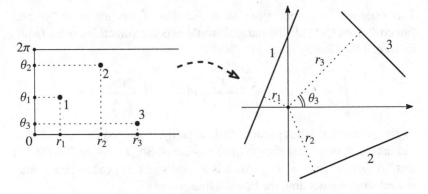

Figure 5.13 Constructing a Poisson line process as a point process in the parameter space.

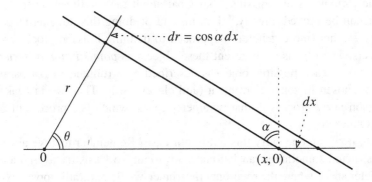

Figure 5.14 Equivalence of the two constructions.

have $\theta = \alpha$ and $dr = \cos \alpha \, dx$, so the probability that there is a line of the process crossing the reference line in the interval $[x, x + dx]$ (with respect to the first coordinate) and having the direction in the interval $[\alpha, \alpha + d\alpha]$ is proportional to $\cos \alpha \, dr \, d\alpha$, as required.

At this point, we prefer to end this discussion and recommend the beautiful book [54] to an interested reader; in particular, that book contains a lot of information about Poisson processes of lines and (hyper)planes.

Finally, here is the general message of this section: it may be possible to construct something which can be naturally called a Poisson process of objects, but the construction may be quite nontrivial. As for the Poisson line process itself, it serves as a supporting example for the previous sentence and as a "get-some-intuition" example for the next chapter,[18] but it is not *directly* connected to anything else in the rest of this book. There is one more reason, however, for its presence here: it is beautiful. As an additional argument in favour of the last affirmation, let us consider the following question: what is the distribution of the direction of a "typical" line of the Poisson line process? Well, it should obviously be uniform (since the process is invariant under rotations). Now, what is the distribution of the direction of a "typical" line intersecting the reference line? This time, by construction, it should obviously obey the cosine law. And here comes the paradox: almost surely *all* lines of the Poisson line process intersect the reference line, so we are talking about the *same* sets of lines! So, what is "the direction of a typical line", after all?

5.3 Exercises

Soft local times (Section 5.1)

Exercise 5.1. Look again at Figure 5.6. Can you find the value of $\sigma(8)$?

Exercise 5.2. Let $(X_i)_{i\geq 1}$ be a Markov chain on a finite set Σ, with transition probabilities $p(x, x')$, initial distribution π_0, and stationary measure π. Let A be a subset of Σ. Prove that for any $n \geq 1$ and $\lambda > 0$ it holds that

$$\mathbb{P}_{\pi_0}[\tau_A \leq n]$$
$$\geq \mathbb{P}_{\pi_0}\left[\xi_0\pi_0(x) + \sum_{j=1}^{n-1} \xi_j p(X_j, x) \geq \lambda\pi(x), \forall x \in \Sigma\right] - e^{-\lambda\pi(A)}, \quad (5.16)$$

where ξ_i are i.i.d. Exp(1) random variables, also independent of the Markov chain X.

Exercise 5.3. Find a nontrivial application of (5.16).

Exercise 5.4. Give a *rigorous* proof of Lemma 5.1 in case Σ is discrete (i.e., finite or countably infinite set).

Exercise 5.5. Let $(X_j)_{j\geq 1}$ be a Markov chain on the state space $\Sigma = \{0, 1\}$, with the following transition probabilities:

$$p(k, k) = 1 - p(k, 1 - k) = \frac{1}{2} + \varepsilon$$

for $k = 0, 1$, where $\varepsilon \in (0, \frac{1}{2})$ is small. Clearly, by symmetry, $(\frac{1}{2}, \frac{1}{2})$ is

[18] In particular, the reader is invited to pay special attention to Exercises 5.12 and 5.13.

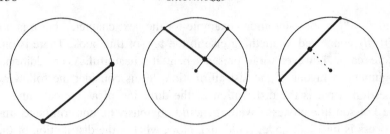

Figure 5.15 Constructing a random chord.

the stationary distribution of this Markov chain. Next, let $(Y_j)_{j\geq 1}$ be a sequence of i.i.d. Bernoulli random variables with success probability $\frac{1}{2}$. Prove that the distance in total variation between the laws of $(X_1,\dots,X_n)$ and $(Y_1,\dots,Y_n)$ is of order ε, uniformly in n.

Poisson processes of objects (Section 5.2)

Exercise 5.6. Let us recall the Bertrand paradox: "what is the probability that a random chord of a circle is longer than the side of the inscribed equilateral triangle?".

The answer, of course, depends on how exactly we decide to choose the random chord. One may consider (at least) *three* apparently natural methods; see Figure 5.15 (from left to right):

  (i) choose two points uniformly and independently, and draw a chord between them;

  (ii) first choose a radius[19] at random, then choose a random point on it (all that uniformly), and then draw the chord perpendicular to the radius through that point;

 (iii) choose a random point *inside the disk* (note that, almost surely, that point will not be the centre), and then draw the unique chord perpendicular to the corresponding radius.

I do not ask you to prove that the probability of the preceding event will be $\frac{1}{3}, \frac{1}{2}$, and $\frac{1}{4}$ respectively for the three aforementioned methods, since it is very easy. Instead, let me ask the following question: how do we find the *right* way to choose a random chord (and therefore resolve the paradox)? One reasonable idea is to consider a Poisson line process (since it is, in a way, the *canonical* random collection of lines on the plane) and *condition* on the fact that only one line intersects the circle, so that this intersection generates the chord. To which of the three methods does it correspond?

[19] i.e., a straight line segment linking the centre to a boundary point

Exercise 5.7. Note that the uniform distribution on a finite set (or a subset of $\mathbb{R}^d$ with finite Lebesgue measure) has the following *characteristic property*: if we condition that the chosen point belongs to a fixed subset, then this conditional distribution is uniform again (on that subset).

Now consider a (smaller) circle which lies fully inside the original circle, and condition that *the* random chord (defined in the *correct* way in the previous exercise) of the bigger circle intersects the smaller one, thus generating a chord in it as well. Verify that this induced random chord has the *correct* distribution.

Exercise 5.8. The preceding method of defining a random chord works for any convex domain. Do you think there is a *right* way of defining a random chord for nonconvex (even nonconnected) domains?

Note that for such a domain one straight line can generate several chords at once.

Exercise 5.9. Explain the paradox in the end of Section 5.2.

Exercise 5.10. Argue that the paradox has a lot to do with the motivating example of Section 5.1; in fact, show how one can generate the Poisson line process using the "strip" representation in two ways (with reference line, and without).

Exercise 5.11 (Random billiards). A particle moves with constant speed inside some (connected, but not necessarily simply connected) domain $\mathcal{D}$. When it hits the boundary, it is reflected in random direction according to the *cosine law*[20] (i.e., with density proportional to the cosine of the angle with the normal vector), independently of the incoming direction, and keeping the absolute value of its speed. Let $X_t \in \mathcal{D}$ be the location of the process at time t, and $V_t \in [0, 2\pi)$ be the corresponding direction; $\xi_n \in \partial\mathcal{D}$, $n = 0, 1, 2, \ldots$ are the points where the process hits the boundary, as shown in Figure 5.16.

Prove that

- the stationary measure of the random walk ξ_n is uniform on $\partial\mathcal{D}$;
- the stationary measure of the process (X_t, V_t) is the product of uniform measures on $\mathcal{D}$ and $[0, 2\pi)$.

Observe that this result holds for any (reasonable) domain $\mathcal{D}$!
The d-dimensional version of this process appeared in [88] under the

---

[20] Recall the construction of the Poisson line process via a reference line.

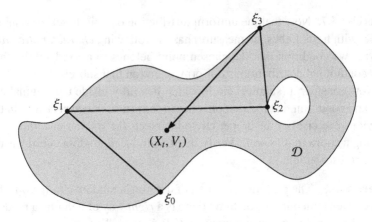

Figure 5.16 Random billiard (starting on the boundary of $\mathcal{D}$).

name of "running shake-and-bake algorithm", and was subsequently studied in [20, 21, 22]. For some physical motivation for the cosine reflection
law, see e.g. [27] and references therein.

Exercise 5.12. Sometimes, instead of defining a Poisson process of infinite objects "as a whole", it is easier to define its image inside a finite
"window". This is not the case for the Poisson line processes,[21] but one
can still do it. Let $A \subset \mathbb{R}^2$ be a convex domain. Prove that the following
procedure defines a Poisson line process as seen in A: take a Poisson point
process on ∂A, and then, independently for each of its points, trace a ray
(pointing inside the domain) according to the cosine law.

Then, prove *directly* (i.e., forget about the Poisson line process in $\mathbb{R}^2$ for
now) that the preceding procedure is *consistent*: if $B \subset A$ is convex too,
then the restriction of the process in A to B has the *correct* law (i.e., the
same as if we took a Poisson point process on ∂B with the same intensity,
traced rays from each of its points, and traced independent rays with the
cosine law); see Figure 5.17.

Exercise 5.13. Now, let us consider two nonintersecting domains $A_1, A_2 \subset \mathbb{R}^2$, and abbreviate $r = \max\{\mathrm{diam}(A_1), \mathrm{diam}(A_2)\}$, $s = \mathrm{dist}(A_1, A_2)$. Consider a two-dimensional Poisson line process with rate λ. It is quite clear
that the restrictions of this process on A_1 and A_2 are *not* independent; just
look at Figure 5.18. However, in the case $s \gg r$ one still can *decouple*
them. Let H_1 and H_2 be two events *supported* on A_1 and A_2. This means

---

[21] I mean, it is not *easier*.

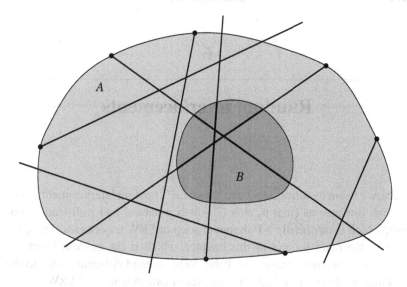

Figure 5.17 Poisson line process seen in a finite convex set.

Figure 5.18 Poisson line process seen in two disjoint sets; if a straight line intersects A_1 the way shown in the picture, it *must* intersect A_2.

that, informally speaking, the occurrence of the event H_k is determined by the configuration seen on A_k, for $k = 1, 2$. Prove that, for some positive constant C we have

$$\left| \mathbb{P}[H_1 \cap H_2] - \mathbb{P}[H_1]\mathbb{P}[H_2] \right| \leq \frac{C\lambda r}{s}. \tag{5.17}$$

Exercise 5.14. How would you define a Poisson line process in higher dimensions? Also, what about the Poisson *plane* process in $\mathbb{R}^d$, $d \geq 3$, etc.?

Exercise 5.15. Find the expected value of the orthogonal projection of the unit cube on a randomly oriented plane.

6

Random interlacements

Here, we begin by introducing the "classical"[1] random interlacement model in high dimensions (that is, $d \geq 3$) – it is a natural and well-studied object, which is informally a Poissonian soup of SRW trajectories. Then, we pass to the main subject of this chapter, which is the random interlacement model in two dimensions. It *has* to be defined differently, due to the fact that, by Theorem 1.1, the trajectories of two-dimensional SRW are a.s. space-filling. It turns out that the *right* way to define this model is to use the trajectories of conditioned SRW of Chapter 4. We then discuss various properties of two-dimensional random interlacements.

6.1 Introduction: random interlacements in higher dimensions

As a warm-up, let us solve Exercises 3.22 and 3.23: we want to prove that for any finite $A \subset B$

$$\mathbb{P}_{\mathrm{hm}_B}[\tau_A < \infty] = \frac{\mathrm{cap}(A)}{\mathrm{cap}(B)}, \tag{6.1}$$

and for all $y \in \partial A$

$$\mathbb{P}_{\mathrm{hm}_B}[S_{\tau_A} = y \mid \tau_A < \infty] = \mathrm{hm}_A(y). \tag{6.2}$$

To do this, write for $x \notin B$

$$\mathbb{P}_x[\tau_A < \infty, S_{\tau_A} = y]$$

(to enter A, have to pass through B)

$$= \sum_{z \in \partial B} \mathbb{P}_x[\tau_A < \infty, S_{\tau_A} = y, S_{\tau_B} = z]$$

$$= \sum_{z \in \partial B} \mathbb{P}_x[\tau_A < \infty, S_{\tau_A} = y, \tau_B < \infty, S_{\tau_B} = z]$$

---

[1] By now.

142

(by the strong Markov property)

$$= \sum_{z \in \partial B} \mathbb{P}_x[\tau_B < \infty, S_{\tau_B} = z] \mathbb{P}_z[\tau_A < \infty, S_{\tau_A} = y].$$

Now let us divide both sides by $G(x)$ and send x to infinity. For the left-hand side, we have by Proposition 3.4 and Theorem 3.8

$$\frac{\mathbb{P}_x[\tau_A < \infty, S_{\tau_A} = y]}{G(x)} = \frac{\mathbb{P}_x[S_{\tau_A} = y \mid \tau_A < \infty]}{G(x)} \mathbb{P}_x[\tau_A < \infty]$$

$$\to \operatorname{cap}(A) \operatorname{hm}_A(y) \quad \text{as } x \to \infty,$$

and, likewise, the right-hand side will become

$$\sum_{z \in \partial B} \frac{\mathbb{P}_x[\tau_B < \infty, S_{\tau_B} = z]}{G(x)} \mathbb{P}_z[\tau_A < \infty, S_{\tau_A} = y]$$

(as $x \to \infty$)

$$\to \sum_{z \in \partial B} \operatorname{cap}(B) \operatorname{hm}_B(z) \mathbb{P}_z[\tau_A < \infty, S_{\tau_A} = y]$$

$$= \operatorname{cap}(B) \mathbb{P}_{\operatorname{hm}_B}[\tau_A < \infty, S_{\tau_A} = y]$$

$$= \operatorname{cap}(B) \mathbb{P}_{\operatorname{hm}_B}[\tau_A < \infty] \mathbb{P}_{\operatorname{hm}_B}[S_{\tau_A} = y \mid \tau_A < \infty],$$

which means that

$$\operatorname{hm}_A(y) = \frac{\operatorname{cap}(B) \mathbb{P}_{\operatorname{hm}_B}[\tau_A < \infty]}{\operatorname{cap}(A)} \times \mathbb{P}_{\operatorname{hm}_B}[S_{\tau_A} = y \mid \tau_A < \infty]. \tag{6.3}$$

Note the following elementary fact: if μ and ν are *probability* measures such that $\mu = c\nu$, then $c = 1$ (and therefore $\mu = \nu$). So, since (6.3) holds for all $y \in \partial A$, this gives us (6.1) and (6.2) at once.

Let us now go back to the main subject of this section. Random interlacements were introduced by Sznitman in [93], motivated by the problem of disconnection of the discrete torus $\mathbb{Z}_n^d := \mathbb{Z}^d / n\mathbb{Z}^d$ by the trace of simple random walk, in dimension 3 or higher. Detailed accounts can be found in the survey [17] and in the recent books [41, 96]. Loosely speaking, the model of random interlacements in $\mathbb{Z}^d$, $d \geq 3$, is a stationary Poissonian soup of (transient) doubly infinite simple random walk trajectories on the integer lattice.

Let us make a stop here and try to *understand* the previous sentence. We want to figure out how a Poisson process of (bi-infinite) SRW's trajectories should look like. It may be not clear at this point how to approach this question in the whole space at once, but let us try to fix a finite $B \subset \mathbb{Z}^d$ and think about how the part of that process which "touches" B should be.

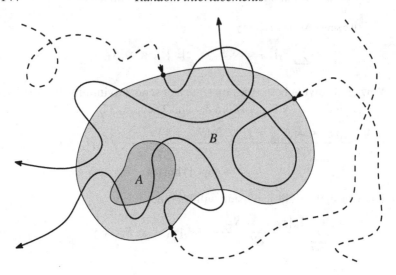

Figure 6.1 A Poisson process of SRW trajectories "restricted"
on B.

First, as mentioned in the previous chapter, that Poisson process of trajectories should be thought of as a Poisson *point* process in another space
(a "space of trajectories"). Since the set of trajectories that intersect B is a
subset of that space, the number of them (i.e., the number of *points* in that
subset) should have Poisson distribution with some parameter λ_B. Since
these trajectories "come from the infinity", they should choose the entrance
location to B according to hm_B; also, as usually happens in the Poissonian
context, they should be independent between each other.

So, for now, we have a Poisson(λ_B) number of independent SRW's trajectories that enter B at random sites chosen according to hm_B and then
continue walking; but how should this parameter λ_B depend on B? To understand this, let us consider a situation depicted in Figure 6.1: take some A
such that $A \subset B$, so the trajectories that intersect A constitute a subset of
those that intersect B. It is here that (6.1) and (6.2) come in handy: by (6.2),
the trajectories that enter A do so via hm_A (as they should), and, by (6.1),
their number has Poisson distribution with parameter $\lambda_B \times \frac{\text{cap}(A)}{\text{cap}(B)}$. Since, on
the other hand, that parameter should not depend on B at all (that is, λ_B must
be equal to $u\,\text{cap}(B)$ for some constant $u > 0$), we arrive to the following
description of the random interlacement process with rate u on $A \subset \mathbb{Z}^d$:

- take Poisson($u\,\text{cap}(A)$) particles and place them on ∂A independently,
  according to hm_A;

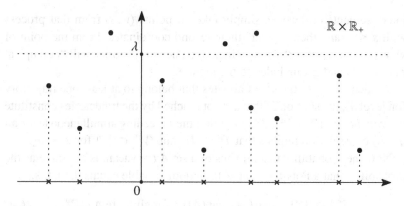

Figure 6.2 A simultaneous construction of one-dimensional Poisson processes: take the points with second coordinate at most λ and project them to obtain a one-dimensional Poisson process with rate λ.

- let them walk.

By the author's experience, the preceding description is the one which is effectively used in most cases when reasoning about random interlacements. Still, it is also important to know that it is possible to define the random interlacement process in the whole $\mathbb{Z}^d$ at once, as a Poisson point process in the space of doubly infinite SRW trajectories. We refer to theorem 1.1 of [93] and also the aforementioned books for the precise construction. The author opted for not reproducing it here since, by his experience, it is at this point that a few students generally disappear from the course and never come back. (Nevertheless, we will briefly summarize a more general construction of random interlacements on transient weighted graphs in the next Section 6.2.) What is important to have in mind, though, is that it is possible to construct the random interlacement process for all $u \geq 0$ *simultaneously*. The idea is exactly the same as in the basic case one-dimensional Poisson processes: all such processes with constant rate[2] can be constructed at once as projections of a Poisson process with rate 1 in $\mathbb{R} \times \mathbb{R}_+$, as in Figure 6.2. Here, the idea is the same: instead of just considering a Poisson point process on a (properly defined) trajectory space W^* for a fixed value of $u > 0$, we rather consider a Poisson point process on $W^* \times \mathbb{R}_+$. This way, the points of the latter process will be of the form (w, s), where w is a trajectory and s is a positive number (the "level" of the trajectory). To obtain the random

---

[2] In fact, even with variable rate.

interlacements on level u, simply take all points (w, s) from that process with $s \leq u$, and then "forget" their second coordinates. From the point of view of a finite set $A \subset \mathbb{Z}^d$, as u increases, new trajectories will "pop up" at rate cap(A) and get included to the picture.

We denote by $\mathcal{I}^u$ the set of all sites that belong to at least one trajectory (on level u); the sites of $\mathbb{Z}^d$ that are not touched by the trajectories constitute the *vacant set* $\mathcal{V}^u = \mathbb{Z}^d \setminus \mathcal{I}^u$. Note that the preceding simultaneous (for all $u > 0$) construction implies that $\mathcal{I}^{u_1} \subset \mathcal{I}^{u_2}$ and $\mathcal{V}^{u_2} \subset \mathcal{V}^{u_1}$ for $u_1 < u_2$.

Now the probability that all sites of a set A are vacant is the same as the probability that a Poisson(u cap(A)) random variable equals 0, that is,

$$\mathbb{P}[A \subset \mathcal{V}^u] = \exp\left(-u\operatorname{cap}(A)\right) \quad \text{for all finite } A \subset \mathbb{Z}^d. \quad (6.4)$$

It is interesting to note that the law of the vacant set $\mathcal{V}^u$ is uniquely characterized by the set of identities (6.4); see theorem 9.2.XII of [28].

As previously mentioned, one of the motivations for studying random interlacements is their relationship with SRW on the torus $\mathbb{Z}_n^d := \mathbb{Z}^d / n\mathbb{Z}^d$. First, it was shown in [9] that the local picture left by the SRW in a *small* subset of $\mathbb{Z}_n^d$ up to time un^d is close in distribution to that of rate-u random interlacements. Then, in [18] it was established that the same is true for a *macroscopic* box in $\mathbb{Z}_n^d$, which was then used to show a sharp phase transition for the diameter of the component of the vacant set on the torus containing a given site.

From now on, we assume that the reader is familiar[3] at least with the definition and the basic facts about the Bernoulli (site) percolation model on $\mathbb{Z}^d$.

We now recall some known properties of the interlacement and the vacant sets; the following list is of course incomplete. For the interlacement set $\mathcal{I}^u$, it holds that:

- $\mathbb{P}[\mathcal{I}^u$ is connected for all $u > 0] = 1$, i.e., the interlacement set is connected *simultaneously* for all $u > 0$ (theorem 1.5 of [16]).

- If we consider a graph whose vertices are the trajectories and there is an edge between two vertices if and only if the corresponding trajectories intersect, then the diameter of this graph equals $\lceil d/2 \rceil$ a.s. (that is, from each site of $\mathcal{I}^u$ it is possible to walk to any other site of $\mathcal{I}^u$, "changing" the underlying trajectories at most $\lceil d/2 \rceil$ times), as shown in [81, 83]. Some further results in this direction can be found in [57].

---

[3] It is clearly a very reasonable assumption to make, but let us still mention e.g. the books [11, 51].

- The *graph distance* (a.k.a. *chemical distance*) on $\mathcal{I}^u$ is *essentially* equivalent to Euclidean distance (except for some "local holes"), see theorem 1.3 of [16].

- SRW on $\mathcal{I}^u$ essentially behaves as the usual SRW [82, 89].

- On the other hand, a random walk with drift on $\mathcal{I}^u$ in three dimensions *always* has zero (even subpolynomial) speed (theorem 1.3 of [47]); this is in sharp contrast with the random walk with drift on the infinite cluster of Bernoulli percolation (somewhat surprisingly, this random walk is ballistic for small values of the drift, and has zero speed for large values of the drift; see [10]). It is conjectured, though, that a random walk with drift on $\mathcal{I}^u$ in dimensions $d \geq 4$ has similar behaviour to that of the drifted random walk on Bernoulli percolation cluster.

While the interlacement set is always connected, this is not so for the vacant set: for example, it is not difficult to convince oneself that it should contain infinitely many "isolated islands". The most important questions, though, are related to the percolative properties of the vacant set: does it contain an infinite cluster, what is the probability of existence of a long path in this set, and so on. The following are some of the known results:

- There is a critical value $u^* \in (0, \infty)$ (which depends also on dimension) such that, for all $u < u^*$, $\mathcal{V}^u$ contains an infinite connected subset a.s., and, for all $u > u^*$, all connected components of $\mathcal{V}^u$ are a.s. finite (see [90]). It is not known (and constitutes a *very* difficult open problem) what happens for $u = u^*$.

- If the infinite vacant cluster exists, then it is unique, as shown in [99].

- At least if u is small enough, the infinite vacant cluster is "well behaved" (i.e., grosso modo, has similar properties to those of the infinite cluster of Bernoulli percolation); see [40, 42, 43, 84, 89, 100].

- There is $u^{**} \in [u^*, \infty)$ such that, for any $u > u^{**}$, the probability that the origin is connected to x within $\mathcal{V}^u$ decays exponentially (with a logarithmic correction for $d = 3$) in $\|x\|$; see [79], theorem 3.1. It is conjectured that $u^* = u^{**}$, but, again, so far it is unclear how to prove this. Observe that this would be an analogue of Menshikov's theorem [69] in Bernoulli percolation.

Compared to the Bernoulli percolation model, the main difficulty in studying the vacant set of random interlacements is that strong (polynomial) correlations are present there. Indeed, note that, by (3.15) and (3.16) for all $x, y \in \mathbb{Z}^d$ it follows from (6.4) that

$$\mathbb{P}[x \in \mathcal{V}^u] = \exp\left(-\frac{u}{G(0)}\right),\tag{6.5}$$

$$\mathbb{P}[\{x, y\} \subset \mathcal{V}^u] = \exp\left(-\frac{2u}{G(0) + G(x - y)}\right).\tag{6.6}$$

Let us now calculate the covariance and the correlation of $\mathbf{1}\{x \in \mathcal{V}^u\}$ and $\mathbf{1}\{y \in \mathcal{V}^u\}$:

$$\mathrm{cov}(\mathbf{1}\{x \in \mathcal{V}^u\}, \mathbf{1}\{y \in \mathcal{V}^u\})$$

$$= \exp\left(-\frac{2u}{G(0) + G(x - y)}\right) - \exp\left(-\frac{2u}{G(0)}\right)$$

(by the way, notice that $\mathrm{cov}(\mathbf{1}\{x \in \mathcal{V}^u\}, \mathbf{1}\{y \in \mathcal{V}^u\}) > 0$) for all $x \neq y$

$$= \exp\left(-\frac{2u}{G(0)}\right)\left(\exp\left(2u\frac{G(x - y)}{G(0) + G(x - y)}\right) - 1\right)$$

(as $x - y \to \infty$)

$$= \frac{2u}{G^2(0)} \exp\left(-\frac{2u}{G(0)}\right) \times \frac{\gamma_d}{\|x - y\|^{d-2}}(1 + O(\|x - y\|^{-(2 \wedge (d-2))})).\tag{6.7}$$

Therefore,

$$\mathrm{corr}(\mathbf{1}\{x \in \mathcal{V}^u\}, \mathbf{1}\{y \in \mathcal{V}^u\})$$

$$= \frac{2u e^{-u/G(0)}}{G^2(0)(1 - e^{-u/G(0)})} \times \frac{\gamma_d}{\|x - y\|^{d-2}}(1 + O(\|x - y\|^{-(2 \wedge (d-2))})).\tag{6.8}$$

That is, when u is fixed, both the covariance and the correlation of $\mathbf{1}\{x \in \mathcal{V}^u\}$ and $\mathbf{1}\{y \in \mathcal{V}^u\}$ have the same order of polynomial decay (with respect to the distance between x and y) as Green's function. It is also interesting to notice that

$$\frac{u e^{-u/G(0)}}{(1 - e^{-u/G(0)})} \sim \begin{cases} u e^{-u/G(0)}, & \text{as } u \to \infty, \\ G(0), & \text{as } u \to 0. \end{cases}$$

That is, the correlation decreases in u as $u \to \infty$ because, intuitively, there are many trajectories that may contribute. The correlation "stabilizes" as $u \to 0$ because the dependence is "embedded" to just one trajectory that may commute between x and y.

This was for the one-site set correlations, but when one considers two nonintersecting (and, usually, *distant*) sets $A_1, A_2 \subset \mathbb{Z}^d$, things start to become even more complicated. Assume that the diameters of these sets are smaller or equal to r and that they are at distance s from each other. Suppose also that we are given two functions $f_1: \{0, 1\}^{A_1} \to [0, 1]$ and

$f_2 \colon \{0, 1\}^{A_2} \to [0, 1]$ that depend only on the configuration of the random interlacements inside the sets A_1 and A_2 respectively. In (2.15) of [93], it was established that

$$\mathrm{Cov}(f_1, f_2) \le c_d u \frac{\mathrm{cap}(A_1)\,\mathrm{cap}(A_2)}{s^{d-2}} \le c'_d u \Big(\frac{r^2}{s}\Big)^{d-2}; \tag{6.9}$$

see also lemma 2.1 of [6]. For this estimate to make sense, we need that $r^2 \ll s$, that is, the sets should be quite far away from each other relative to their sizes. It is remarkable that, even with these restrictions, the preceding inequality can be used to obtain highly non-trivial results. The estimate in (6.9) is not optimal and can be improved to some extent, as was shown in [29]. Still, it is an open problem to find the *correct* maximal order of decay of the covariance in (6.9). Since the random interlacement model is actually closely related to the so-called *Gaussian Free Field* (see [94, 95]) for which that maximal order of decay is known to be $O(\frac{\sqrt{\mathrm{cap}(A_1)\,\mathrm{cap}(A_2)}}{s^{d-2}})$, as shown in [78], it would be natural to conjecture that the same should hold for random interlacements. Still, the decay of correlations would remain polynomial and the estimate only can be useful when $r \ll s$.

As we just mentioned, even these weaker estimates can be very useful; one still naturally looks for more, though. It turns out that one can obtain estimates with much smaller error term if one is allowed to change the level parameter u (this is called *sprinkling*). To explain what it means, we first need to introduce the notion of *monotone* (increasing or decreasing) event/function.

Definition 6.1. Let $A \subset \mathbb{Z}^d$; for configurations of zeros and ones $\{\eta \colon A \to \{0, 1\}\} = \{0, 1\}^A$, there is a natural partial order: $\eta \le \zeta$ whenever $\eta(x) \le \zeta(x)$ for all $x \in A$. We then say that a function $f \colon \{0, 1\}^A \to \mathbb{R}$ is *increasing* if $f(\eta) \ge f(\zeta)$ for all $\eta \ge \zeta$; a function g is *decreasing* if $(-g)$ is increasing. We say that an *event* determined by a configuration is increasing/decreasing if its indicator function is as well.

If we denote by $\mathbb{P}^u$ and $\mathbb{E}^u$ the probability and the expectation with respect to the level-u random interlacements (using the convention that the state of x is 0 if it is vacant and 1 if it is occupied), from the "simultaneous" construction of random interlacements it then immediately follows that $\mathbb{E}^u f$ and $\mathbb{P}^u[H]$ are increasing (respectively, decreasing) functions of u for any increasing (respectively, decreasing) function f or event H.

It turns out that most *interesting* events (and functions) are monotone. For example:

- the proportion of occupied (vacant) sites in A;

- existence of path of occupied (vacant) sites in A that connects a fixed site $x_0 \in A$ to ∂A;
- existence of a crossing (is some suitable sense) by occupied (vacant) sites in A;
- and so on.

An important fact about monotone events is that, in the case of random interlacements the so-called Harris–FKG inequality is still true (as it is for the Bernoulli percolation model). Before formulating it, note that a monotone function depending on $A \subset \mathbb{Z}^d$ can be also seen as a monotone function depending on $A' \subset \mathbb{Z}^d$, where $A \subset A'$ (i.e., it would not depend on the configuration on $A' \setminus A$, but we can still formally regard it as a function $\{0, 1\}^{A'} \to \mathbb{R}$). The Harris–FKG inequality proved in [98] asserts that, given two functions $f_{1,2} : \{0, 1\}^A \to \mathbb{R}$ which are both increasing or both decreasing, it holds that

$$\mathbb{E}^u f_1 f_2 \geq \mathbb{E}^u f_1 \, \mathbb{E}^u f_2, \tag{6.10}$$

that is, the increasing functions (events) are nonnegatively correlated. It then follows that (6.10) also holds when the increasing functions $f_{1,2}$ depend on *disjoint* sets $A_{1,2}$, since both functions can be seen as functions on $\{0, 1\}^{A_1 \cup A_2}$; so $\mathrm{Cov}(f_1, f_2) \geq 0$ for monotone (both increasing or both decreasing) functions $f_{1,2}$. The Harris–FKG inequality can be often useful; however, it works only in one direction. In the other direction, we can aim for inequalities of the form $\mathbb{E}^u f_1 f_2 \leq \mathbb{E}^u f_1 \, \mathbb{E}^u f_2 +$ error term, which follow from (6.9) and the like. However, in any case, that error term has to be at least polynomial (with respect to the distance between the sets), since, as shown in (6.7), it is already polynomial for the one-site sets. In many situations, this "slow" decay can complicate the things quite considerably. Here comes the key idea (known as *sprinkling*): if, for some (small) δ, we substitute $\mathbb{E}^u f_1 \, \mathbb{E}^u f_2$ with $\mathbb{E}^{u+\delta} f_1 \, \mathbb{E}^{u+\delta} f_2$ in the right-hand side, the error term can be made considerably smaller! Indeed, the following result was proved in [79] (theorem 1.1):

Theorem 6.2. *Let A_1, A_2 be two nonintersecting subsets of $\mathbb{Z}^d$, with at least one of them being finite. Let $s = \mathrm{dist}(A_1, A_2)$ and r be the minimum of their diameters. Then for all $u > 0$ and $\varepsilon \in (0, 1)$, we have*

(i) for any increasing functions $f_1 : \{0, 1\}^{A_1} \to [0, 1]$ and $f_2 : \{0, 1\}^{A_2} \to [0, 1]$,

$$\mathbb{E}^u f_1 f_2 \leq \mathbb{E}^{(1+\varepsilon)u} f_1 \mathbb{E}^{(1+\varepsilon)u} f_2 + c'(r + s)^d \exp(-c'' \varepsilon^2 u s^{d-2}); \tag{6.11}$$

(ii) for any decreasing functions $f_1: \{0,1\}^{A_1} \to [0,1]$ and $f_2: \{0,1\}^{A_2} \to [0,1]$,

$$\mathbb{E}^u f_1 f_2 \le \mathbb{E}^{(1-\varepsilon)u} f_1 \mathbb{E}^{(1-\varepsilon)u} f_2 + c'(r+s)^d \exp(-c'' \varepsilon^2 u s^{d-2}) \quad (6.12)$$

(here and in the next two theorems the positive constants c' and c'' only depend on dimension).

Observe that these inequalities are operational even when the two sets are close (relative to the size of the smallest of them). They are consequences of the next result (theorem 2.1 of [79]):

Theorem 6.3. *Let A_1, A_2, s and r be as in Theorem 6.2. Then for all $u > 0$ and $\varepsilon \in (0,1)$ there exists a coupling between $(I^u)_{u \ge 0}$ and two independent random interlacement processes, $(I^u_1)_{u \ge 0}$ and $(I^u_2)_{u \ge 0}$, such that*

$$\mathbb{P}[I_k^{u(1-\varepsilon)} \cap A_k \subset I^u \cap A_k \subset I_k^{u(1+\varepsilon)}, \ k=1,2]$$
$$\ge 1 - c'(r+s)^d \exp(-c'' \varepsilon^2 u s^{d-2}). \quad (6.13)$$

In words, it shows us that there is a way to decouple the intersection of the interlacement set I^u with two disjoint subsets A_1 and A_2 of $\mathbb{Z}^d$. Namely, we couple the original interlacement process I^u with two *independent* interlacements processes I^u_1 and I^u_2 in such a way that I^u restricted on A_k is "close" to I^u_k, for $k = 1, 2$, with probability rapidly going to 1 as the distance between the sets increases.

In fact, what was really proven in [79] (sadly, without explicitly formulating it) is an even more general result, which would imply, for example, a similar decoupling for local times of random interlacements (local time at $x \in \mathbb{Z}^d$ is the aggregate visit count of x by *all* the trajectories). To be able to formulate that general result, first, recall the notion of excursions from Chapter 4 (see Figure 4.2): if $B = \partial B'$ for some finite B' such that $A \subset B'$, an excursion between A and B is a finite piece of nearest-neighbour trajectory which begins at a site of ∂A and ends on the first visit of B. For A, B as before and $u > 0$, let us denote by $\mathcal{T}^u_{A,B}$ the set of excursions between A and B generated by the random interlacement's trajectories up to level u. Also, we say that a (finite) set V *separates* A_1 from A_2, if any nearest-neighbour path which goes from A_1 to A_2 has to pass through V.

Now, also for reference purposes, we state the result that was *really* proved in [79]:

Theorem 6.4. *Let A_1, A_2, s, and r be as in Theorems 6.2 and 6.3. Then there exists a set V that separates A_1 from A_2 and such that for all $u > 0$ and $\varepsilon \in (0,1)$ there exists a coupling between the set of excursions $\mathcal{T}^u_{A_1 \cup A_2, V}$ and*

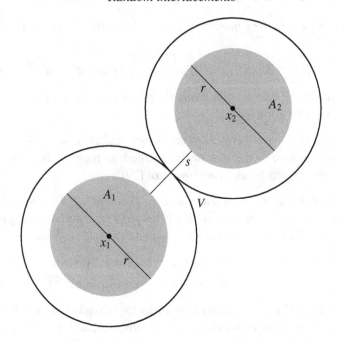

Figure 6.3 The simplified setup for the proof of Theorem 6.4.

*two independent families of sets of excursions (i.e., related to independent
random interlacement processes)* $(\mathcal{T}^u_{A_1,V})_{u\geq 0}$ *and* $(\mathcal{T}^u_{A_2,V})_{u\geq 0}$, *such that*

$$\mathbb{P}[(\mathcal{T}^{(1-\varepsilon)u}_{A_1,V} \cup \mathcal{T}^{(1-\varepsilon)u}_{A_2,V}) \subset \mathcal{T}^u_{A_1\cup A_2,V} \subset (\mathcal{T}^{(1+\varepsilon)u}_{A_1,V} \cup \mathcal{T}^{(1+\varepsilon)u}_{A_2,V})]$$
$$\geq 1 - c'(r+s)^d \exp(-c''\varepsilon^2 u s^{d-2}). \tag{6.14}$$

One of the possible choices of the separating set V is

$$V = \partial\{x \in \mathbb{Z}^d : \mathrm{dist}(x, A_1) \leq s/2\}.$$

The proof of the preceding results is essentially an application of the soft
local times technique. In the following, we sketch the proof of Theorem 6.4
in the simpler case when A_1 and A_2 are (discrete) balls of equal sizes; also,
we will assume that $r \geq s \geq 1$ (that is, $A_{1,2}$ are not too far away from each
other). Let $A_{1,2} = \mathsf{B}(x_{1,2}, r/2)$, and assume that $s := \|x_1 - x_2\| - r > 3$; in this
case, it is easy to see that $\mathrm{diam}(A_{1,2}) = r + O(1)$ and $\mathrm{dist}(A_1, A_2) = s + O(1)$.
Let us define

$$V = \partial\mathsf{B}(x_1, \tfrac{r}{2} + \tfrac{s}{2}) \cup \mathsf{B}(x_2, \tfrac{r}{2} + \tfrac{s}{2})$$

(see Figure 6.3). Clearly, V separates A_1 from A_2.

Next, the idea is the following: using the method of soft local times, we *simultaneously* construct excursions between V and A (where A is one of the sets A_1, A_2, $A_1 \cup A_2$) on level u in the case of $A_1 \cup A_2$ and on levels $u \pm \varepsilon$ in the case of A_1 and A_2. Notice the following important fact: given the initial site of an excursion, its subsequent evolution is (conditionally) independent from the past; due to this feature, we can avoid using (a bit complicated) space of all excursions as the "base space" in the soft local times method's setup. That is, an equivalent way[4] of performing the soft local times construction of these excursions is the following. Let Δ be the "cemetery state", corresponding to the situation when the random walk's trajectory goes to infinity without touching A anymore. First, we take a Poisson point process on $(\partial A \cup \Delta) \times \mathbb{R}_+$ (with simply the counting measure on $\partial A \cup \Delta$ as the base measure), with *marks* attached to each point (see Figure 6.4): each mark is a random excursion started there in the case the chosen point is a site of ∂A (i.e., it is an SRW started at $x \in \partial A$ and observed till the first hitting of V), and it is just void ("null excursion") in the case when the chosen point happened to be Δ. We will successively obtain the excursions corresponding to the Poisson($v \, \text{cap}(A)$) trajectories by picking up the corresponding marks. Let $Z_n^{(0)}, Z_n^{(\ell)} \in V \cup \{\Delta\}$ be the "first and last sites of the nth excursion", in the sense that they are really its first and last sites in the case when the excursion actually took place, and they are both Δ in the case the walk went to infinity without touching A anymore; we formally define $Z_0^{(0)} = Z_0^{(\ell)} = \Delta$. With $\mathcal{F}_n$ being the sigma-algebra generated by the first n excursions, the densities $g(\cdot \mid \mathcal{F}_n)$ are defined in the following way:

- first, $g(\cdot \mid \mathcal{F}_{n-1}, Z_{n-1}^{(\ell)} = \Delta) = \text{hm}_A(\cdot)$ (that is, if the previous trajectory escaped to infinity, we start a new excursion from a site chosen according to the harmonic measure);
- next, given that the previous excursion ended in $z \in V$, we have $g(x \mid \mathcal{F}_{n-1}, Z_{n-1}^{(\ell)} = z) = \mathbb{P}_z[\tau_A < \infty, S_{\tau_A} = x]$ for $x \in \partial A$ and $g(\Delta \mid \mathcal{F}_{n-1}, Z_{n-1}^{(\ell)} = \Delta) = \mathbb{P}_z[\tau_A = \infty]$.

When using this construction to obtain the excursions of $\mathcal{T}_{A,V}^v$, we need to deal with soft local times with random[5] index: indeed, the total number of trajectories which give rise to those excursions has Poisson($v \, \text{cap}(A)$) distribution. In the following, we will denote that soft local time by $\mathcal{G}_v^A(\cdot)$ (since V does not vary, we keep it out of this notation).

---

[4] We leave it as an exercise to show that it is indeed so.
[5] With a compound Poisson distribution.

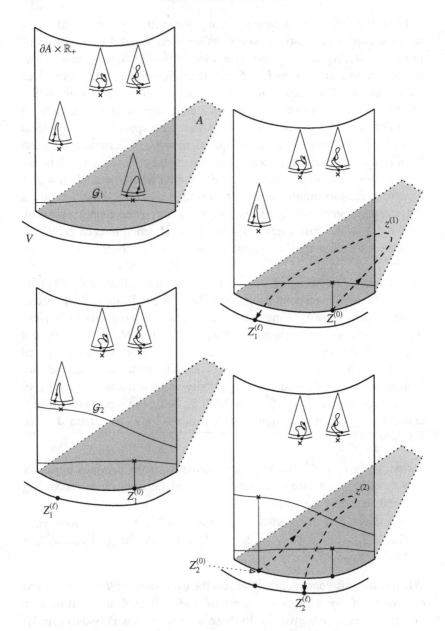

Figure 6.4 On the construction of random interlacements on the set A; the points of Σ_A are substituted by sites in $\partial A \times \mathbb{R}_+$ with marks representing the corresponding trajectories; the state Δ is not in the figure.

For $z \in \partial(A_1 \cup A_2)$, define $\iota(z) = 1$ when $z \in \partial A_1$ and $\iota(z) = 2$ in the case $z \in \partial A_2$. Then, as discussed in Section 5.1, we need to show that for every $u > 0$ and $\varepsilon \in (0, 1)$ there exists a coupling between $\mathcal{T}^u_{A_1 \cup A_2, V}$ and two independent processes $(\mathcal{T}^v_{A_1, V})_{v \geq 0}$ and $(\mathcal{T}^v_{A_2, V})_{v \geq 0}$ such that

$$\mathbb{P}[(\mathcal{T}^{(1-\varepsilon)u}_{A_1, V} \cup \mathcal{T}^{(1-\varepsilon)u}_{A_2, V}) \subset \mathcal{T}^u_{A_1 \cup A_2, V} \subset (\mathcal{T}^{(1+\varepsilon)u}_{A_1, V} \cup \mathcal{T}^{(1+\varepsilon)u}_{A_2, V})]$$
$$\geq \mathbb{P}[\mathcal{G}^{A_{\iota(z)}}_{(1-\varepsilon)u}(z) \leq \mathcal{G}^{A_1 \cup A_2}_u(z) \leq \mathcal{G}^{A_{\iota(z)}}_{(1+\varepsilon)u}(z) \text{ for all } z \in \partial(A_1 \cup A_2)]. \quad (6.15)$$

The main idea for constructing such a coupling is the usual one with soft local times: use the same Poisson points to construct all processes we need. In this concrete case, we just construct the up-to-level u excursions between $A_1 \cup A_2$ and V, and then construct *separately* the excursion processes between A_1 and V and between A_2 and V, using the same points with the same marks (clearly, the last two processes are independent since they "live" on nonintersecting sets). Now, the crucial observation is that, for $k = 1, 2$,

$$\mathbb{E}\mathcal{G}^{A_k}_v(z) = \mathbb{E}\mathcal{G}^{A_1 \cup A_2}_v(z) \qquad \text{for all } v > 0 \text{ and } z \in \partial A_k. \quad (6.16)$$

To see that (6.16) indeed holds, denote first by $L^v_{A, V}(z)$ the number of random interlacement's excursions between A and V up to level v that have $z \in \partial A$ as their first site. It is then clear that, for any $z \in \partial A_k$, $L^v_{A_k, V}(z) = L^v_{A_1 \cup A_2, V}(z)$, $k = 1, 2$ (because, since V separates A_1 from A_2, those are just the same excursions). Therefore, Proposition 5.4 (the one that says that expected local time equals expected soft local time) implies $\mathbb{E}\mathcal{G}^{A_k}_v(z) = \mathbb{E}L^v_{A_k, V}(z) = \mathbb{E}L^v_{A_1 \cup A_2, V}(z) = \mathbb{E}\mathcal{G}^{A_1 \cup A_2}_v(z)$. So, let us denote (clearly, the expectations in (6.16) are linear in v) $\varphi(z) = v^{-1}\mathbb{E}\mathcal{G}^{A_1 \cup A_2}_v(z) = v^{-1}\mathbb{E}\mathcal{G}^{A_k}_v(z)$ for $k = 1, 2$.

Now it is straightforward to check that $(1 - \varepsilon)(1 + \frac{\varepsilon}{3}) \leq (1 - \frac{\varepsilon}{3})$ and $(1 + \frac{\varepsilon}{3}) \leq (1 + \varepsilon)(1 - \frac{\varepsilon}{3})$ for all $\varepsilon \in [0, 1]$. Therefore, with the union bound, (6.15) implies that (see Figure 6.5)

$$\mathbb{P}[(\mathcal{T}^{(1-\varepsilon)u}_{A_1, V} \cup \mathcal{T}^{(1-\varepsilon)u}_{A_2, V}) \subset \mathcal{T}^u_{A_1 \cup A_2, V} \subset (\mathcal{T}^{(1+\varepsilon)u}_{A_1, V} \cup \mathcal{T}^{(1+\varepsilon)u}_{A_2, V})]$$
$$\geq 1 - \sum_{\substack{(v,A)=((1\pm\varepsilon)u, A_1), \\ ((1\pm\varepsilon)u, A_2), (u, A_1 \cup A_2)}} \mathbb{P}[|\mathcal{G}^A_v(z) - v\varphi(z)| \geq \tfrac{\varepsilon}{3}v\varphi(z) \text{ for some } z \in \partial A]. \quad (6.17)$$

To deal with the right-hand side of (6.17), the plan is to first obtain an upper bound on the corresponding probability for a *fixed* $z \in \partial A$. Recall that, by definition, the number Θ^A_v of trajectories on level v that touch A has Poisson($v \operatorname{cap}(A)$) distribution. Define $\eta_0 := 0$, $\eta_k = \min\{j > \eta_{k-1} : Z^{(\ell)}_j = $

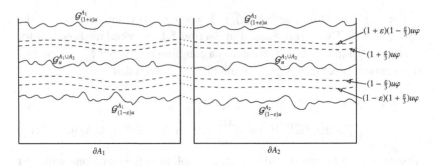

Figure 6.5 "Separating" the soft local times of the excursion
processes.

$\Delta\}$ to be the indices of the "void excursions", and let

$$F_k^A(z) = \sum_{i=\eta_{k-1}+1}^{\eta_k} g(z \mid \mathcal{F}_{i-1})$$

be the contribution of the kth trajectory to the soft local time in $z \in \partial A$.
Then the soft local time $\mathcal{G}_v^A(z)$ is a sum of i.i.d. random variables:

$$\mathcal{G}_v^A(z) = \sum_{k=1}^{\Theta_v^A} F_k^A(z). \tag{6.18}$$

So, we need to obtain large-deviation estimates on that sum. For that we
need to know (or at least estimate) the first and second moments of these
random variables, and we need also some tail estimates on them.

Recall that A stands for one of the sets A_1, A_2, $A_1 \cup A_2$. For $z \in \partial A$, let
us denote $\pi^A(z) = \mathbb{E}F_1^A(z)$. We need the following fact:

Lemma 6.5. *It holds that*

(i) $\pi^A(z) \asymp s^{-1} r^{-(d-2)}$;
(ii) $\mathbb{E}(F_1^A(z))^2 \lesssim s^{-d} r^{-(d-2)}$.

Sketch of the proof. This is a particular case of lemma 6.2 of [79], which
is proved using also a general upper bound on second moments of soft
local times (theorem 4.8 of [79]). Here, we only informally explain why (i)
and (ii) should be true. For this, we need the following technical fact (stated
below as Exercise 6.2): if $z \in \partial A$ and $x \in \mathbb{Z}^d \setminus A$ is at distance of order s from
both A and z, then $\mathbb{P}_x[S_{\tau_A} = z]$ is of order $s^{-(d-1)}$ (this is because, basically,
the number of natural "entrance point candidates" is of order s^{d-1}).

To show part (i), we recall Proposition 5.4, which asserts that the expectation of the soft local time is equal to the expectation of the "real" local time. In our case, this amounts to estimating the expectation of the number of excursions (generated by one trajectory) which enter A exactly at z. Now, the probability that this trajectory ever hits $B(z, s/3)$ is of order $\frac{s^{d-2}}{r^{d-2}}$ (this can be obtained e.g. using Exercise 3.24). Then, if that happened, the probability of entering through z will be at most $s^{-(d-1)}$; also, starting from any site of V, the probability of ever visiting $B(z, s/3)$ is bounded away from 1. This gives an upper bound of order $\frac{s^{d-2}}{r^{d-2}} \times s^{-(d-1)} = s^{-1}r^{-(d-2)}$; a lower bound of the same order may be obtained if one notices that a probability of ever visiting a part of V which is at distance of order s from z is still of order $\frac{s^{d-2}}{r^{d-2}}$; (6.65) then would imply the lower bound we need.

As for part (ii), the intuition is the following (think of the case $s \ll r$):

- there is only a significant contribution to the soft local time when the last site $Z^{(\ell)}$ of the previous excursion is at distance of order s from z;
- as before, the chance that the trajectory ever gets at distance of order s from z is $O(\frac{s^{d-2}}{r^{d-2}})$;
- if the preceding happened, then the contribution to the soft local time will be of order $s^{-(d-1)}$, so, the second moment should be of order $\frac{s^{d-2}}{r^{d-2}} \times (s^{-(d-1)})^2 = s^{-d}r^{-(d-2)}$.

$\square$

Next, we need the following large deviation bound for $F_1^A(z)$:

Lemma 6.6. *For $A = A_1, A_2, A_1 \cup A_2$ and V as before, for any $h \geq 2$ we have*

$$\mathbb{P}[F_1^A(z) > h\gamma s^{-(d-1)}] \leq c' \frac{s^{d-2}}{r^{d-2}} \exp(-c''h), \qquad (6.19)$$

for all $z \in \partial A$.

Sketch of the proof. This is a particular case of lemma 6.3 of [79], which by its turn follows from a general tail estimate on the soft local time provided by theorem 4.9 of [79]. Here, we again only explain informally why (6.19) should hold. In fact, the idea is really similar to the argument for estimating the second moment that we have just seen. First, the trajectory has to come to distance of order s from z; this gives rise to the factor $\frac{s^{d-2}}{r^{d-2}}$ in (6.19). Then, after each excursion, there is a uniformly positive chance that the trajectory leaves that s-neighbourhood of z for good (and essentially stops "contributing" to the soft local time); this explains the exponent in (6.19). $\square$

The rest of the proof is a standard Chernoff bound type argument for sums of i.i.d. random variables: the soft local times we are interested in (recall (6.17)) are so by (6.18). From now on and till the end of the proof, we adopt a rather loose attitude towards the use of constants: "c" will stand for a generic positive constant (depending only on dimension) whose value may change from line to line. We refer to the last two pages of [79] for a cleaner treatment of this argument. For $A = A_1, A_2, A_1 \cup A_2$ and $x \in \partial A$, let $\psi_A^x(\lambda) = \mathbb{E}e^{\lambda F_1^A(z)}$ be the moment generating function of $F_1^A(z)$. It is elementary to obtain that $e^t - 1 \leq t + t^2$ for all $t \in [0, 1]$. Using this observation, we write for $\lambda = O(s^{d-1})$

$$\psi_A^x(\lambda) - 1$$
$$= \mathbb{E}(e^{\lambda F_1^A(z)} - 1)\mathbf{1}\{\lambda F_1^A(z) \leq 1\} + \mathbb{E}(e^{\lambda F_1^A(z)} - 1)\mathbf{1}\{\lambda F_1^A(z) > 1\}$$
$$\leq \mathbb{E}(\lambda F_1^A(z) + \lambda^2(F_1^A(z))^2) + \mathbb{E}e^{\lambda F_1^A(z)}\mathbf{1}\{F_1^A(z) > \lambda^{-1}\}$$
$$\leq \lambda\pi^A(z) + c\lambda^2 s^{-d} r^{-(d-2)} + \lambda \int_{\lambda^{-1}}^{\infty} e^{\lambda y}\mathbb{P}[F_1^A(z) > y]\,dy$$
$$\leq \lambda\pi^A(z) + c\lambda^2 s^{-d} r^{-(d-2)} + c\lambda s^{d-2} r^{-(d-2)} \int_{\lambda^{-1}}^{\infty} \exp(-cs^{d-1}y)\,dy$$
$$\leq \lambda\pi^A(z) + c\lambda^2 s^{-d} r^{-(d-2)} + cs^{-1} r^{-(d-2)}\lambda \exp(-c\lambda^{-1} s^{d-1})$$
$$\leq \lambda\pi^A(z) + c\lambda^2 s^{-d} r^{-(d-2)}, \tag{6.20}$$

where we used Lemma 6.5 (ii) and Lemma 6.6. Analogously, since $e^{-t}-1 \leq -t + t^2$ for *all* $t > 0$, we obtain for $\lambda \geq 0$

$$\psi_A^x(-\lambda) - 1 \leq -\lambda\pi^A(z) + c\lambda^2 s^{-d} r^{-(d-2)} \tag{6.21}$$

(here we do not need the large deviation bound of Lemma 6.6).

Observe that, if $(Y_k, k \geq 1)$ are i.i.d. random variables with a common moment generating function ψ and Θ is an independent Poisson random variable with parameter θ, then $\mathbb{E}\exp\left(\lambda \sum_{k=1}^{\Theta} Y_k\right) = \exp\left(\theta(\psi(\lambda) - 1)\right)$. So, using (6.20) and Lemma 6.5 (ii), we write for any $\delta > 0$, $z \in \Sigma$ and $x = X_0(z)$,

$$\mathbb{E}[\mathcal{G}_v^A(z) \geq (1 + \delta)v\,\mathrm{cap}(A)\pi^A(z)]$$
$$= \mathbb{E}\left[\sum_{k=1}^{\Theta_v^A} F_k^A(z) \geq (1 + \delta)v\,\mathrm{cap}(A)\pi^A(z)\right]$$

$$\leq \frac{\mathbb{E}\exp\left(\lambda\sum_{k=1}^{\Theta_v^A} F_k^A(z)\right)}{\exp\left(\lambda(1+\delta)v\,\mathrm{cap}(A)\pi^A(z)\right)}$$

$$= \exp\left(-\lambda(1+\delta)v\,\mathrm{cap}(A)\pi^A(z) + v\,\mathrm{cap}(A)(\psi(\lambda)-1)\right)$$

$$\leq \exp\left(-(\lambda\delta v\,\mathrm{cap}(A)\pi^A(z) - c\lambda^2 v s^{-d})\right)$$

$$\leq \exp\left(-(c\lambda\delta v s^{-1} - c\lambda^2 v s^{-d})\right),$$

and, analogously, with (6.21) instead of (6.20), one can obtain

$$\mathbb{E}\left[\mathcal{G}_v^A(z) \leq (1-\delta)v\,\mathrm{cap}(A)\pi^A(z)\right] \leq \exp\left(-(c\lambda\delta v s^{-1} - c\lambda^2 v s^{-d})\right).$$

We choose $\lambda = c\delta s^{d-1}$ with small enough c and then use also the union bound (clearly, the cardinality of ∂A is at most $O(r^d)$) to obtain that

$$\mathbb{P}\left[(1-\delta)v\,\mathrm{cap}(A)\pi^A(z) \leq \mathcal{G}_v^A(z) \leq (1+\delta)v\,\mathrm{cap}(A)\pi^A(z) \text{ for all } z\right]$$
$$\geq 1 - cr^d\exp\left(-c\delta^2 v s^{d-2}\right). \quad (6.22)$$

Using (6.22) with $\delta = \frac{\varepsilon}{3}$ and $u, (1-\varepsilon)u, (1+\varepsilon)u$ on the place of $\hat{u}$ together with (6.17), we conclude the proof of Theorem 6.4 in this particular case. □

The "classical" random interlacement model in $d \geq 3$ remains a very interesting and active research topic; in particular, there are further partial advances on decoupling inequalities, both with and without sprinkling [3, 29]. However, let us remind ourselves that this book is supposed to be mostly about two dimensions, and end this section here.

6.2 The two-dimensional case

At first glance, the title of this section seems to be meaningless, just because (as we remember from Chapter 2) even a single trajectory of two-dimensional simple random walk a.s. visits all sites of $\mathbb{Z}^2$; so the vacant set would be always empty and the local time will be infinite at all sites. Let us not rush to conclusions, though, and think about the following. As mentioned in the previous section, "classical" random interlacements in dimensions $d \geq 3$ are related to the simple random walk on the discrete torus $\mathbb{Z}_n^d$: the picture left by the SRW up to time un^d on a "small" window is well approximated by random interlacements on level u. It is true, though, that one can hardly expect something of this sort in two dimensions, at least for a window of fixed size and fixed location. Indeed, consider a fixed $A \subset \mathbb{Z}_n^2$ (that is, the size of A is constant while n grows); since SRW on $\mathbb{Z}_n^2$ inherits a "local recurrence" property of two-dimensional SRW, if we know that one site of A was visited by a given (fixed and large) time, then it is also likely that all sites of A were visited. So, we will typically see either "A full" or "A empty", and anything more interesting than that is unlikely to happen.

Now it is time for a digression: surprisingly enough, in some sense, SRW on $\mathbb{Z}_n^2$ is a more complicated model than SRW on $\mathbb{Z}_n^d$ for $d \geq 3$ – there are quite a few questions that are considerably easier in higher dimensions. For example, it was only relatively recently proved in [33] that the expected *cover time* of $\mathbb{Z}_n^2$ is equal to $\frac{4}{\pi} n^2 \ln^2 n$ plus[6] terms of smaller order. A question which (to the best of the author's knowledge) is still not fully understood in two dimensions is what the last uncovered sites of the torus look like (in higher dimensions, it is known that they are essentially independent and uniformly distributed; see [7, 73]). For example: what can we say about the last two uncovered sites on $\mathbb{Z}_n^2$? Should they be "almost neighbours" with high probability, or at least with probability uniformly bounded away from 0? More generally, what can be said about the probability distribution of the (graph) distance between them?

The author tried to look at this last question a few years ago, reasoning in the following way. As we just mentioned, in higher dimensions the last uncovered sites are (almost) independent. So, one can imagine that, if we *condition* that a given site is *vacant* (uncovered) in the regime when only a few sites should still remain vacant, the probability that a neighbour of that site is vacant as well should be close to 0. It then looks like a reasonable idea to try calculating this sort of probability also in two dimensions.

Let us do it, at least on a heuristic level.

Denote $t_\alpha := \frac{4\alpha}{\pi} n^2 \ln^2 n$ for $\alpha > 0$, so that t_α is roughly the expected cover time of the torus multiplied by α. Also, let $U_m^{(n)} \subset \mathbb{Z}_n^2$ be the (random) set of sites which remain uncovered by time m by the SRW which starts from a site chosen uniformly at random. We single out one vertex of the torus and call it 0; let us identify the torus with the (discrete) square on $\mathbb{Z}^2$ placed in such a way that 0 is (roughly) at its centre. Let us consider the excursions of the random walk between $\partial B(\frac{n}{3\ln n})$ and $\partial B(n/3)$ up to time t_α. It is possible to show (using, for example, lemma 3.2 of [34]) that the number E_α of these excursions is concentrated around $\frac{2\alpha \ln^2 n}{\ln \ln n}$, with deviation probabilities of subpolynomial order. This is for unconditional probabilities, but, since the probability of the event $\{0 \in U_{t_\alpha}^{(n)}\}$ is only polynomially small (actually, it is $n^{-2\alpha + o(1)}$), the same holds for the deviation probabilities conditioned on

---

[6]　In fact, minus; see [1, 8].

this event. So, let us just assume for now that the number of the excursions is *exactly* $\frac{2\alpha \ln^2 n}{\ln \ln n}$ and see where will it lead us.

Now, consider a (fixed) set A such that $0 \in A$ (we think of A "being present simultaneously" on the torus and on the plane; of course, we assume that it is fully inside the square of size n centred at 0). Then (3.41) and (3.63) imply that

- the probability that an excursion hits the origin is roughly $\frac{\ln \ln n}{\ln(n/3)}$;
- provided that $\text{cap}(A) \ll \ln n$, the probability that an excursion hits the set A is roughly $\frac{\ln \ln n}{\ln(n/3)}(1 + \frac{\pi \, \text{cap}(A)}{2\ln(n/3)})$.

So, the *conditional* probability p_* that an excursion does not hit A given that it does not hit the origin is

$$p_* \approx \frac{1 - \frac{\ln \ln n}{\ln(n/3)}(1 + \frac{\pi \, \text{cap}(A)}{2\ln(n/3)})}{1 - \frac{\ln \ln n}{\ln(n/3)}} \approx 1 - \frac{\pi \ln \ln n}{2\ln^2 n}\, \text{cap}(A),$$

and then we obtain

$$\mathbb{P}[A \subset U_{t_\alpha}^{(n)} \mid 0 \in U_{t_\alpha}^{(n)}] \approx p_*^{E_\alpha}$$

$$\approx \left(1 - \frac{\pi \ln \ln n}{2\ln^2 n}\, \text{cap}(A)\right)^{\frac{2\alpha \ln^2 n}{\ln \ln n}}$$

$$\approx \exp\left(-\pi\alpha \, \text{cap}(A)\right). \tag{6.23}$$

The preceding argument is, of course, *very* heuristic (we comment more on that after Theorem 6.11), but let us accept it for now, and look at the expression (6.23). We have seen a very similar thing quite recently, namely, in the formula (6.4) for the probability that a finite subset of $\mathbb{Z}^d$ is vacant for the (classical) random interlacement model. This *suggests* that there may be some natural random interlacement model in two dimensions as well. This is indeed the case: later, we will see that it is possible to construct it in such a way that

$$\mathbb{P}[A \subset \mathcal{V}^\alpha] = \exp\left(-\pi\alpha \, \text{cap}(A)\right) \tag{6.24}$$

holds *for all finite sets containing the origin*[7] (the factor π in the exponent is just for convenience, as explained later). Naturally enough (recall that the left-hand side of (6.23) is a conditional probability), to build the interlacements we use trajectories of simple random walks conditioned on never hitting the origin. Of course, the law of the vacant set is no longer

---

[7] Note, however, that it is clearly not possible for (6.24) to hold for *every* finite A, since the two-dimensional capacity of one-site sets equals 0.

translationally invariant, but we show that it has the property of *conditional translation invariance*, cf. Theorem 6.8 below.

Now it is time to write some formalities.[8] Here, we will use the general construction of random interlacements on a transient weighted graph introduced in [98]. In the following few lines, we briefly summarize this construction. Let W be the space of all doubly infinite nearest-neighbour transient trajectories in $\mathbb{Z}^2$,

$$W = \{\varrho = (\varrho_k)_{k \in \mathbb{Z}} : \varrho_k \sim \varrho_{k+1} \text{ for all } k;$$
$$\text{the set } \{m \in \mathbb{Z} : \varrho_m = y\} \text{ is finite for all } y \in \mathbb{Z}^2\}.$$

We say that ϱ and ϱ' are equivalent if they coincide after a time shift, i.e., $\varrho \sim \varrho'$ when there exists k such that $\varrho_{m+k} = \varrho_m$ for all $m \in \mathbb{Z}$. Then, let $W^* = W/\sim$ be the space of such trajectories modulo time shift, and define χ^* to be the canonical projection from W to W^*. For a finite $A \subset \mathbb{Z}^2$, let W_A be the set of trajectories in W that intersect A, and we write W_A^* for the image of W_A under χ^*. One then constructs the random interlacements as the Poisson point process on $W^* \times \mathbb{R}^+$ with the intensity measure $\nu \otimes du$, where ν is described in the following way. It is the unique sigma-finite measure on W^* such that for every finite A

$$1_{W_A^*} \cdot \nu = \chi^* \circ Q_A,$$

where the finite measure Q_A on W_A with total mass $\widehat{\text{cap}}(A)$ is determined by the following equality:

$$Q_A[(\varrho_k)_{k \geq 1} \in F, \varrho_0 = x, (\varrho_{-k})_{k \geq 1} \in H]$$
$$= \widehat{\text{cap}}(A) \, \widehat{\text{hm}}_A(x) \cdot \mathbb{P}_x[\widehat{S} \in F] \cdot \mathbb{P}_x[\widehat{S} \in H \mid \hat{\tau}_A^+ = \infty]. \tag{6.25}$$

The existence and uniqueness of ν were shown in theorem 2.1 of [98].

Definition 6.7. For a configuration $\sum_\lambda \delta_{(w_\lambda^*, u_\lambda)}$ of the preceding Poisson process, the *process of random interlacements at level* α (which will be referred to as RI(α)) is defined as the set of trajectories with a label less than or equal to $\pi\alpha$, i.e.,

$$\sum_{\lambda : u_\lambda \leq \pi\alpha} \delta_{w_\lambda^*}.$$

Observe that this definition is somewhat unconventional (we used $\pi\alpha$ instead of just α, as one would do in higher dimensions), but we will see later that it is quite reasonable in two dimensions, since the formulas become generally cleaner.

---

[8] And leave absorbing these formalities as an exercise.

It is important to have in mind the following *constructive* description of random interlacements at level α observed on a finite set $A \subset \mathbb{Z}^2$ (this is, in fact, implied by the preceding formal construction; note the terms $\widehat{\mathrm{hm}}_A(x)\mathbb{P}_x[\widehat{S} \in F]$ in (6.25)). Namely,

- take a Poisson($\pi\alpha\,\widehat{\mathrm{cap}}(A)$) number of particles;
- place these particles on the boundary of A independently, with distribution $\widehat{\mathrm{hm}}_A$;
- let the particles perform independent $\widehat{S}$-random walks (since $\widehat{S}$ is transient, each walk only leaves a finite trace on A).

We then immediately see that

$$\mathbb{P}[A \subset \mathcal{V}^\alpha] = \exp\left(-\pi\alpha\,\widehat{\mathrm{cap}}(A)\right) \qquad (6.26)$$

for *all* finite $A \subset \mathbb{Z}^2$ (indeed, A is vacant if and only if no trajectory hits it); Theorem 4.12 then implies that

$$\mathbb{P}[A \subset \mathcal{V}^\alpha] = \exp\left(-\pi\alpha\,\mathrm{cap}(A)\right)$$

for all finite subsets A of $\mathbb{Z}^2$ containing the origin.

It is also worth mentioning that the Harris–FKG inequality still holds for two-dimensional random interlacements, cf. theorem 3.1 of [98].

The *vacant set* at level α,

$$\mathcal{V}^\alpha = \mathbb{Z}^2 \setminus \bigcup_{\lambda:u_\lambda \le \pi\alpha} \omega_\lambda^*(\mathbb{Z}),$$

is the set of lattice points not covered by the random interlacement. It contains the origin by definition. In Figure 6.6, we present a simulation[9] of the vacant set for different values of the parameter α.

As mentioned before, the law of two-dimensional random interlacements is not translationally invariant, although it is of course invariant with respect to reflections/rotations of $\mathbb{Z}^2$ that preserve the origin. Let us describe some other basic properties of two-dimensional random interlacements:

Theorem 6.8. *(i) For any $\alpha > 0$, $x \in \mathbb{Z}^2$, $A \subset \mathbb{Z}^2$, it holds that*

$$\mathbb{P}[A \subset \mathcal{V}^\alpha \mid x \in \mathcal{V}^\alpha] = \mathbb{P}[-A + x \subset \mathcal{V}^\alpha \mid x \in \mathcal{V}^\alpha]. \qquad (6.27)$$

More generally, for all $\alpha > 0$, $x \in \mathbb{Z}^2 \setminus \{0\}$, $A \subset \mathbb{Z}^2$, and any lattice isometry M exchanging 0 and x, we have

$$\mathbb{P}[A \subset \mathcal{V}^\alpha \mid x \in \mathcal{V}^\alpha] = \mathbb{P}[MA \subset \mathcal{V}^\alpha \mid x \in \mathcal{V}^\alpha]. \qquad (6.28)$$

---

[9] Many thanks to Darcy Camargo for doing it!

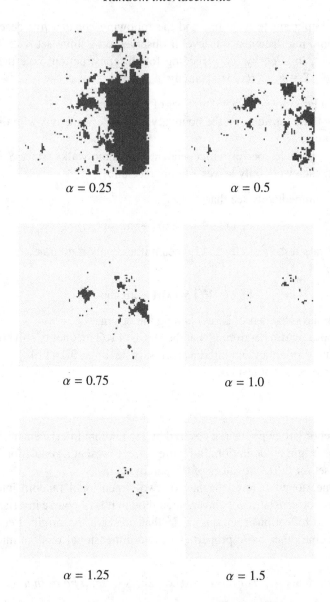

$\alpha = 0.25$ $\alpha = 0.5$

$\alpha = 0.75$ $\alpha = 1.0$

$\alpha = 1.25$ $\alpha = 1.5$

Figure 6.6 A realization of the vacant set (dark) of $RI(\alpha)$ on a square 101×101 (centred at the origin) for different values of α. For $\alpha = 1.5$ the only vacant site is the origin. Also, note that we see the same neighbourhoods of the origin for $\alpha = 1$ and $\alpha = 1.25$; this is not surprising since just a few new walks enter the picture when the rate increases by a small amount.

(ii) *With γ' from (3.36) we have*

$$\mathbb{P}[x \in \mathcal{V}^\alpha] = \exp\left(-\pi\alpha\frac{a(x)}{2}\right)$$

$$= \exp\left(-\frac{\gamma'\pi\alpha}{2}\right)\|x\|^{-\alpha}(1 + O(\|x\|^{-2})). \tag{6.29}$$

(iii) *Let A be a finite subset of $\mathbb{Z}^2$ such that $0 \in A$ and $|A| \geq 2$, and denote $r = 1 + \max_{z \in A} \|z\|$; let $x \in \mathbb{Z}^2$ be such that $\|x\| \geq 2r$. Then we have*

$$\mathbb{P}[A \subset \mathcal{V}^\alpha \mid x \in \mathcal{V}^\alpha] = \exp\left(-\frac{\pi\alpha}{4}\operatorname{cap}(A)\frac{1 + O(\frac{r}{\|x\|\ln r})}{1 - \frac{\operatorname{cap}(A)}{2a(x)} + O(\frac{r}{\|x\|\ln\|x\|})}\right). \tag{6.30}$$

(iv) *Let $x, y \neq 0$, $x \neq y$, and assume that, as $s := \|x\| \to \infty$, $\ln\|y\| \sim \ln s$ and $\ln\|x - y\| \sim \beta\ln s$ with some $\beta \in [0, 1]$. Then we have*

$$\mathbb{P}[\{x, y\} \subset \mathcal{V}^\alpha] = s^{-\frac{4\alpha}{4-\beta} + o(1)}. \tag{6.31}$$

(v) *Assume that $\ln\|x\| \sim \ln s$, $\ln r \sim \beta\ln s$ with $\beta < 1$. Then, as $s \to \infty$,*

$$\mathbb{P}[B(x, r) \subset \mathcal{V}^\alpha] = s^{-\frac{2\alpha}{2-\beta} + o(1)}. \tag{6.32}$$

These results invite a few comments.

Remark 6.9. Regarding Theorem 6.8:

(a) The statement in (i) describes an invariance property given that a point is vacant. We refer to it as the conditional stationarity of two-dimensional random interlacements.

(b) We can interpret (iii) as follows: the conditional law of RI(α) given that a distant[10] site x is vacant, is similar – near the origin – to the unconditional law of RI($\alpha/4$). Combined with (i), the similarity holds near x as well. Observe that the relation (4.21) leads to the following heuristic explanation for Theorem 6.8 (iii) (in the case when A is fixed and $\|x\| \to \infty$). Since the probability of hitting a distant site is about $1/2$, by conditioning that this distant site is vacant, we essentially throw away three quarters of the trajectories that pass through a neighbourhood of the origin: indeed, the double-infinite trajectory has to avoid this distant site two times, before and after reaching that neighbourhood.

(c) By symmetry, the conclusion of (iv) remains the same in the situation when $\ln\|x\|$, $\ln\|x - y\| \sim \ln s$ and $\ln\|y\| \sim \beta\ln s$.

---

[10] In fact, a *very* distant – the term $\frac{\operatorname{cap}(A)}{2a(x)}$ needs to be small.

Proof of (i) and (ii)   To prove (i), observe that

$$\text{cap}\,(\{0, x\} \cup A) = \text{cap}\,(\{0, x\} \cup (-A + x))$$

because of symmetry. For the second statement in (i), note that, for $A' = \{0, x\} \cup A$, it holds that

$$\text{cap}\,(A') = \text{cap}\,(MA') = \text{cap}\,(\{0, x\} \cup MA).$$

Item (ii) follows from the fact that $\text{cap}\,(\{0, x\}) = \frac{1}{2}a(x)$ (recall (3.65)) together with (3.36). □

We postpone the proof of other parts of Theorem 6.8, since it requires some estimates for capacities of various kinds of sets. We now turn to estimates on the cardinality of the vacant set.

Theorem 6.10.   *(i) We have, with γ' from (3.36)*

$$\mathbb{E}|\mathcal{V}^{\alpha} \cap B(r)| \sim \begin{cases} \frac{2\pi}{2-\alpha}\exp\left(-\frac{\gamma'\pi\alpha}{2}\right)r^{2-\alpha}, & \text{for } \alpha < 2, \\[2mm] 2\pi\exp\left(-\frac{\gamma'\pi\alpha}{2}\right)\ln r, & \text{for } \alpha = 2, \\[2mm] const, & \text{for } \alpha > 2, \end{cases}$$

as $r \to \infty$.

(ii) For $\alpha > 1$, it holds that $\mathcal{V}^{\alpha}$ is finite a.s. Moreover, $\mathbb{P}[\mathcal{V}^{\alpha} = \{0\}] > 0$ and $\mathbb{P}[\mathcal{V}^{\alpha} = \{0\}] \to 1$ as $\alpha \to \infty$.

(iii) For $\alpha \in (0, 1]$, we have $|\mathcal{V}^{\alpha}| = \infty$ a.s. Moreover, for $\alpha \in (0, 1)$, it holds that

$$\mathbb{P}[\mathcal{V}^{\alpha} \cap (B(r) \setminus B(r/2)) = \emptyset] \leq r^{-2(1-\sqrt{\alpha})^2 + o(1)}. \qquad (6.33)$$

It is worth noting that the phase transition at $\alpha = 1$ in (ii) and (iii) corresponds to the cover time of the torus, as shown in Theorem 6.11 below. For that reason, RI(1) is referred to as the *critical* two-dimensional random interlacement model.[11]

Proof of (i) and (ii) (incomplete, in the latter case)   Part (i) readily follows from Theorem 6.8 (ii).

The proof of part (ii) is easy in the case $\alpha > 2$. Indeed, observe first that $\mathbb{E}|\mathcal{V}^{\alpha}| < \infty$ implies that $\mathcal{V}^{\alpha}$ itself is a.s. finite. Also, Theorem 6.8 (ii)

---

[11]   Incidentally, it has a connection to the so-called *Gelfond's constant*
$e^{\pi} = (-1)^{-i} \approx 23.14069\ldots$: the probability that a (fixed) neighbour of the origin is vacant in the critical random interlacement is one divided by square root of Gelfond's constant.

actually implies that $\mathbb{E}|V^\alpha \setminus \{0\}|$ converges to 0 as $\alpha \to \infty$, which implies that $\mathbb{P}[|V^\alpha \setminus \{0\}| = 0] = \mathbb{P}[V^\alpha = \{0\}] \to 1$ by Chebyshev's inequality.

Now let us prove that, in general, $\mathbb{P}[|V^\alpha| < \infty] = 1$ implies that $\mathbb{P}[V^\alpha = \{0\}] > 0$. Indeed, if V^α is a.s. finite, then one can find a sufficiently large R such that $\mathbb{P}[|V^\alpha \cap (\mathbb{Z}^2 \setminus B(R))| = 0] > 0$. Since $\mathbb{P}[x \notin V^\alpha] > 0$ for any $x \neq 0$, the claim that $\mathbb{P}[V^\alpha = \{0\}] > 0$ follows from the Harris–FKG inequality applied to events $\{x \notin V^\alpha\}$, $x \in B(R)$ together with $\{|V^\alpha \cap (\mathbb{Z}^2 \setminus B(R))| = 0\}$ (see also Exercise 6.8). $\qquad\square$

As before, we postpone the proof of part (iii) and the rest of part (ii) of Theorem 6.10, except for the case $\alpha = 1$. The proof that V^1 is a.s. infinite is quite involved and can be found in [23]. Let us only mention that this result may seem somewhat surprising, for the following reason. Recall that the case $\alpha = 1$ corresponds to the leading term in the expression for the cover time of the two-dimensional torus. Then, as already mentioned (cf. [1, 8]), the cover time has a *negative* second-order correction, which could be an evidence in favour of finiteness of V^1 (informally, the "real" all-covering regime should be "just below" $\alpha = 1$). On the other hand, it turns out that local fluctuations of excursion counts overcome that negative correction, thus leading to the previous result.

Let us now give a heuristic explanation about the unusual behaviour of the model for $\alpha \in (1,2)$: in this nontrivial interval, the vacant set is a.s. finite, but its expected size is infinite. The reason for that is the following: the number of $\widehat{S}$-walks that hit $B(r)$ has Poisson law with a rate of order $\ln r$. Thus, decreasing this number by a constant factor (with respect to the expectation) has only a polynomial cost. On the other hand, by doing so, we increase the probability that a site $x \in B(r)$ is vacant for all $x \in B(r)$ at once, which increases the expected size of $V^\alpha \cap B(r)$ by a polynomial factor. It turns out that this effect causes the actual number of uncovered sites in $B(r)$ to be typically of much smaller order than the expected number of uncovered sites there.

In the remaining part of this section, we briefly discuss the relationship of random interlacements with the SRW $(X_n, n \geq 0)$ on the two-dimensional torus $\mathbb{Z}_n^2$. Denote by $\Upsilon_n \colon \mathbb{Z}^2 \to \mathbb{Z}_n^2$,

$$\Upsilon_n(x) = (x_1 \mod n, x_2 \mod n) \quad \text{for } x = (x_1, x_2) \in \mathbb{Z}^2$$

the natural projection modulo n. Then, if the initial position S_0 of a two-dimensional SRW were chosen uniformly at random on any fixed $n \times n$ square, we can write $X_k = \Upsilon_n(S_k)$. Similarly, $B(y, r) \subset \mathbb{Z}_n^2$ is defined by $B(y, r) = \Upsilon_n(B(z, r))$, where $z \in \mathbb{Z}^2$ is such that $\Upsilon_n z = y$. Let us also recall the notations $t_\alpha = \frac{4\alpha}{\pi} n^2 \ln^2 n$ (note that, as mentioned previously, $\alpha = 1$ corresponds to the leading-order term of the expected cover time of the torus), and $U_m^{(n)}$ (which is the set of uncovered sites of $\mathbb{Z}_n^2$ up to time m). In the following theorem, we have that, given that $0 \in \mathbb{Z}_n^2$ is uncovered, the law of the uncovered set around 0 at time t_α is close to that of $\mathrm{RI}(\alpha)$:

Theorem 6.11. *Let $\alpha > 0$ and A be a finite subset of $\mathbb{Z}^2$ such that $0 \in A$. Then we have*

$$\lim_{n\to\infty} \mathbb{P}[\Upsilon_n A \subset U_{t_\alpha}^{(n)} \mid 0 \in U_{t_\alpha}^{(n)}] = \exp(-\pi\alpha \operatorname{cap}(A)). \qquad (6.34)$$

We will not present a rigorous proof of this theorem; let us only mention that it can be proved using a time-dependent version of Doob's transform. It is important to observe that turning the heuristics from the beginning of this section into a proof is not an easy task. The reason for this is that, although E_α is indeed concentrated around $\frac{2\alpha \ln^2 n}{\ln \ln n}$, it is not *concentrated enough*: the probability that 0 is not hit during k excursions, where k varies over the "typical" values of E_α, changes too much. Therefore, in the proof of Theorem 6.11 a different route is needed.

We refer to [87] for a much more in-depth treatment of the relationship between the SRW on the two-dimensional torus and random interlacements. Also, in [2] a model similar to the SRW on the torus was studied (there, the walk takes place on a bounded subset of $\mathbb{Z}^2$, and, upon leaving it, the walk re-enters through a random boundary edge on the next step), also establishing connections with two-dimensional random interlacements.

6.3 Proofs for two-dimensional random interlacements

6.3.1 Excursions and soft local times

In this section, we will develop some tools for dealing with excursions of two-dimensional random walks and/or random interlacements, using the method of soft local times of Section 5.1.

The results that we obtain here will be valid in several contexts: for SRW S on $\mathbb{Z}^2$ or on a torus $\mathbb{Z}_n^2$, and for conditioned SRW $\widehat{S}$. To keep the exposition general, let us denote the underlying process by X.

Assume that we have a finite collection of sets $A_j \subset A_j'$, $j = 1, \ldots, k_0$, where $A_1', \ldots, A_{k_0}'$ are disjoint and $\operatorname{dist}(A_i', A_j') > 1$ for $i \neq j$. Denote also $A = \bigcup_j A_j$ and $A' = \bigcup_j A_j'$; note that we then have $\partial A = \bigcup_j \partial A_j$ and $\partial A' = \bigcup_j \partial A_j'$. Here and in the sequel we denote by $(\mathscr{E}_i^{(j)}, i \geq 1)$ the (complete) excursions (recall Figure 4.2 in Section 4.2) of the *process X* between ∂A_j

and $\partial A'_j$. Observe that these excursions will *appear* in some order in the overall sequence of excursions between ∂A and $\partial A'$; for example, for $k_0 = 2$ we may see $\mathcal{E}x_1^{(2)}, \mathcal{E}x_2^{(2)}, \mathcal{E}x_1^{(1)}, \mathcal{E}x_3^{(2)}$, and so on.

Then suppose that we are given a collection of probability measures $(\nu_j, j = 1, \ldots, k_0)$ with $\text{supp}(\nu_j) = \partial A_j$ (typically, ν_j will be hm_{A_j} or $\widehat{\text{hm}}_{A_j}$, or close to either). Let us now assume that the excursions of the process X were obtained using the soft local time method: we use a *marked* Poisson point process $\{(z_\theta, u_\theta), \theta \in \Theta\}$ with $z_\theta \in \partial A$ and $u_\theta > 0$; the marks at $\{(z_\theta, u_\theta)\}$ are independent excursions starting at z_θ. For $j = 1, \ldots, k_0$ let us then denote by $(\widetilde{\mathcal{E}}x^{(j)})_i, i \geq 1$ the *corresponding*[12] independent excursions with the initial sites chosen according to ν_j and the subsequent evolution according to X (recall that $\mathcal{E}x$'s and $\widetilde{\mathcal{E}}x$'s are actually *the same* excursions, only picked in a different order). For $j = 1, \ldots, k_0$ and $b_2 > b_1 > 0$, define the random variables

$$\mathcal{Z}_j(b_1, b_2) = \#\{\theta \in \Xi : z_\theta \in \partial A_j, b_1 \nu_j(z_\theta) < u_\theta \leq b_2 \nu_j(z_\theta)\}. \quad (6.35)$$

It should be observed that the analysis of the soft local times is considerably simpler in this section than in the proof of Theorem 6.4 (of Section 6.1). This is because here the (conditional) entrance measures to A_j are typically very close to each other (as in (6.36)). That permits us to make *sure* statements about the comparison of the soft local times for different processes in cases when the realization of the Poisson process in $\partial A_j \times \mathbb{R}_+$ is sufficiently well behaved, as e.g. in (6.37). The following is a version of lemma 2.1 of [25]:

Lemma 6.12. *Assume that the probability measures* $(\nu_j, j = 1, \ldots, k_0)$ *are such that for all* $y \in \partial A'$, $x \in \partial A_j$, $j = 1, \ldots, k_0$, *and some* $v \in (0, 1)$, *we have*

$$1 - \frac{v}{3} \leq \frac{\mathbb{P}_y[X_{\tau_A} = x \mid X_{\tau_A} \in A_j]}{\nu_j(x)} \leq 1 + \frac{v}{3}. \quad (6.36)$$

Furthermore, for $m_0 \geq 1$ *define the events*

$$U_j^{m_0} = \{\text{for all } m \geq m_0 \text{ we have } \mathcal{Z}_j(m, (1+v)m) < 2vm$$
$$\text{and } (1-v)m < \mathcal{Z}_j(0, m) < (1+v)m\}. \quad (6.37)$$

Then for all $j = 1, \ldots, k_0$ *and all* $m_0 \geq 1$, *it holds that*

(i) $\mathbb{P}[U_j^{m_0}] \geq 1 - c_1 \exp(-c_2 v m_0)$, *and*

---

[12] I.e., constructed with soft local times using the same realization of the Poisson process.

(ii) on the event $U_j^{m_0}$ we have[13] for all $m \geq m_0$

$$\{\widetilde{\mathcal{E}x}_1^{(j)}, \ldots, \widetilde{\mathcal{E}x}_{(1-v)m}^{(j)}\} \subset \{\mathcal{E}x_1^{(j)}, \ldots, \mathcal{E}x_{(1+3v)m}^{(j)}\},$$

$$\{\mathcal{E}x_1^{(j)}, \ldots, \mathcal{E}x_{(1-v)m}^{(j)}\} \subset \{\widetilde{\mathcal{E}x}_1^{(j)}, \ldots, \widetilde{\mathcal{E}x}_{(1+3v)m}^{(j)}\}.$$

Proof Fix any $j \in \{1, \ldots, k_0\}$ and observe that $\mathcal{Z}_j(b_1, b_2)$ has Poisson distribution with parameter $b_2 - b_1$. It is then straightforward to obtain (i) using usual large deviation bounds.

To prove (ii), fix $k \geq 1$ and let (recall that $\mathcal{G}_k(y)$ denotes the soft local time at y at time k)

$$y_j^{(k)} = \arg\min_{y \in \partial A_j} \frac{\mathcal{G}_k(y)}{\nu_j(y)}$$

(with the convention $0/0 = +\infty$). We then argue that for all $k \geq 1$ we *surely* have

$$\frac{\mathcal{G}_k(y)}{\nu_j(y)} \leq (1+v)\frac{\mathcal{G}_k(y_j^{(k)})}{\nu_j(y_j^{(k)})} \qquad \text{for all } y \in \partial A_j. \tag{6.38}$$

Indeed, by (6.36) we have (denoting by ζ_k the last site of kth excursion, which can also be ∞ for random interlacements when the new trajectory is started)

$$\begin{aligned}
\frac{\mathcal{G}_k(y)}{\nu_j(y)} &= \frac{1}{\nu_j(y)} \sum_{i=1}^{k} \xi_i \mathbb{P}_{\zeta_{i-1}}[X_{\tau_A} = y] \\
&= \sum_{i=1}^{k} \xi_i \frac{\mathbb{P}_{\zeta_{i-1}}[X_{\tau_A} = y \mid X_{\tau_A} \in A_j]}{\nu_j(y)} \mathbb{P}_{\zeta_{i-1}}[X_{\tau_A} \in A_j] \\
&\leq \frac{1+\frac{v}{3}}{1-\frac{v}{3}} \cdot \sum_{i=1}^{k} \xi_i \frac{\mathbb{P}_{\zeta_{i-1}}[X_{\tau_A} = y_j^{(k)} \mid \zeta_{i-1} \in A_j]}{\nu_j(y_j^{(k)})} \mathbb{P}_{\zeta_{i-1}}[X_{\tau_A} \in A_j] \\
&\leq (1+v)\frac{\mathcal{G}_k(y_j^{(k)})}{\nu_j(y_j^{(k)})},
\end{aligned}$$

since $(1 + \frac{v}{3})/(1 - \frac{v}{3}) \leq 1 + v$ for $v \in (0, 1)$.

Let us denote by $\sigma_s^{(j)}$ the position of sth excursion between ∂A_j and $\partial A'_j$ in the sequence of all excursions between ∂A and $\partial A'$. Now let $m \geq m_0$, and abbreviate $\ell = \sigma_{(1-v)m}^{(j)}$. We then have $\mathcal{G}_\ell(y_j^{(\ell)})/\nu_j(y_j^{(\ell)}) \leq m$ (because otherwise, recalling (6.37), we would have more than $(1-v)m$ points of

---

[13] A quick note on indices: e.g. $\mathcal{E}x_s^{(j)}$ is meant to be $\mathcal{E}x_{\lfloor s \rfloor}^{(j)}$ in case s is not an integer.

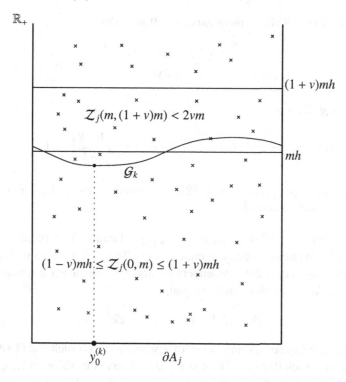

Figure 6.7 On the proof of Lemma 6.12. For simplicity, here we assumed that $v_j \equiv h$ for a positive constant h.

the Poisson process below the graph of $\mathcal{G}_\ell$). So, by (6.38), $\mathcal{G}_\ell(y)/v_j(y) \leq (1+v)m$ for all $y \in \partial A_j$ (see Figure 6.7), which implies that

$$\{\mathcal{E}_{\kappa_1}^{(j)}, \ldots, \mathcal{E}_{\kappa_{(1-v)m}}^{(j)}\} \subset \{\widetilde{\mathcal{E}}_{\kappa_1}^{(j)}, \ldots, \widetilde{\mathcal{E}}_{\kappa_{(1+3v)m}}^{(j)}\}.$$

Analogously, for $\ell' = \sigma_{(1+3v)m}^{(j)}$ we must have $\mathcal{G}_{\ell'}(y_0^{(\ell')})/v_j(y_0^{(\ell')}) \geq m$ (because otherwise $\mathcal{G}_{\ell'}(\cdot)/v_j(\cdot)$ would lie strictly below $(1+v)m$, and we would have $\mathcal{Z}_j(0, (1+v)m) < (1+3v)m)$, so

$$\{\widetilde{\mathcal{E}}_{\kappa_1}^{(j)}, \ldots, \widetilde{\mathcal{E}}_{\kappa_{(1-v)m}}^{(j)}\} \subset \{\mathcal{E}_{\kappa_1}^{(j)}, \ldots, \mathcal{E}_{\kappa_{(1+3v)m}}^{(j)}\}.$$

This concludes the proof of Lemma 6.12.   □

Next, let us obtain the following consequence of Lemma 6.12:

Lemma 6.13. *Let $1 \leq r < R$ be such that $r \geq \frac{R}{\ln^h R}$ for some fixed $h > 0$.*

Then for any $\varphi \in (0, 1)$, there exists $\delta > 0$ such that

$$\sup_{\substack{z \in \partial B(R) \\ y \in \partial B(r)}} \left| \frac{\mathbb{P}_z[S_{\tau_{B(r)}} = y \mid \tau_{B(r)} < \tau^+_{B(R)^c}]}{\mathrm{hm}_{B(r)}(y)} - 1 \right| < \delta \qquad (6.39)$$

then, as $R \to \infty$,

$$\mathbb{P}\left[\exists y \in B(r) \text{ such that } y \notin \widetilde{\mathcal{E}}_i \text{ for all } i \le 2\varphi \frac{\ln^2 R}{\ln \frac{R}{r}}\right] \to 1, \qquad (6.40)$$

where $\widetilde{\mathcal{E}}_1, \widetilde{\mathcal{E}}_2, \widetilde{\mathcal{E}}_3, \ldots$ are i.i.d. SRW excursions between $\partial B(r)$ and $\partial B(R)$ with entrance measure $\mathrm{hm}_{B(r)}$.

Proof Set $n = 3R + 1$ and $k_1 = 2\varphi \frac{\ln^2 R}{\ln R/r}$. Lemma 6.12 (with X being the SRW) implies that one can choose a small enough $\delta > 0$ in such a way that one may couple the independent excursions with the excursion process $\mathcal{E}_1, \mathcal{E}_2, \mathcal{E}_3, \ldots$ of SRW on $\mathbb{Z}_n^2$ so that

$$\{\widetilde{\mathcal{E}}_1, \ldots, \widetilde{\mathcal{E}}_{k_1}\} \subset \{\mathcal{E}_1, \ldots, \mathcal{E}_{(1+\delta')k_1}\}$$

with probability converging to 1 with n, where $\delta' > 0$ is such that $(1+\delta')\varphi < 1$. choose b such that $(1+\delta')\varphi < b < 1$ and observe that theorem 1.2 of [34] implies that a fixed ball with radius at least $\frac{n}{\ln^b n}$ will not be completely covered up to time $\frac{4}{\pi} bn^2 \ln^2 n$ with probability converging to 1. Now we need to control the number of SRW's excursions which take place up to that time. This can be done using lemma 3.2 of [34]: it asserts that there exist $\varepsilon_0 > 0$, $c > 0$ such that if $r < R \le \frac{n}{2}$ and $\varepsilon \le \varepsilon_0$ with $\varepsilon \ge 6c_1(\frac{1}{r} + \frac{r}{R})$, we have for all $x_0 \in \mathbb{Z}_n^2$

$$\mathbb{P}_{x_0}\left[j\text{th excursion is completed before time } (1+\varepsilon)\frac{2n^2 \ln \frac{R}{r}}{\pi} j\right]$$

$$\ge 1 - \exp\left(-\frac{c\varepsilon^2 \ln \frac{R}{r}}{\ln \frac{n}{r}} j\right).$$

This implies that[14]

$$\mathbb{P}[B(r) \text{ is not completely covered by } \{\mathcal{E}_1, \ldots, \mathcal{E}_{(1+\delta')k_1}\}] \to 1$$

as $n \to \infty$, and this completes the proof of (6.40). □

---

[14] Observe that solving $\frac{4}{\pi} bn^2 \ln^2 n = (1+\varepsilon)\frac{2n^2 \ln R/r}{\pi} j$ for j gives $j = \frac{2b}{1+\varepsilon} \frac{\ln^2 n}{\ln R/r}$, and that $\frac{\ln^2 n}{\ln^2 R} \to 1$.

6.3.2 Hitting probabilities

First, we need an estimate for $\hat{g}(y, y + v)$ in the case $\|v\| \ll \|y\|$: as usual, with (3.38) we write

$$
\begin{aligned}
\hat{g}(y, y + v) &= \frac{a(y) + a(y + v) - a(v)}{a(y)a(y + v)} \\
&= \frac{2a(y) - a(v) + O(\frac{\|v\|}{\|y\|})}{a(y)(a(y) + O(\frac{\|v\|}{\|y\|}))} \\
&= \frac{2a(y) - a(v) + O(\frac{\|v\|}{\|y\|})}{a^2(y)}.
\end{aligned} \tag{6.41}
$$

Let us state a couple of estimates, for the probability of (not) hitting a given set (which is, typically, far away from the origin), or, more specifically, a disk. The following is a cleaner version of lemma 3.7 of [26]:

Lemma 6.14. *Assume that $x \notin B(y, r)$ and $\|y\| > 2r \geq 1$.*

(i) We have

$$
\mathbb{P}_x[\hat{\tau}_{B(y,r)} < \infty] = \frac{a(y)(a(y) + a(x) - a(x - y))}{a(x)(2a(y) - a(r) + O(r^{-1} + \frac{r}{\|y\|}))}. \tag{6.42}
$$

(ii) Consider now a finite nonempty set $A \subset \mathbb{Z}^2 \setminus \{0\}$ such that $y \in A$, denote $r = 1 + \max_{z \in A} \|y - z\|$, and assume that $\|y\| > 2r$. Then for any x such that $\|x - y\| \geq 5(r + 1)$, it holds that

$$
\mathbb{P}_x[\hat{\tau}_A < \infty] = \frac{a(y)(a(y) + a(x) - a(x - y))}{a(x)(2a(y) - \mathrm{cap}(A) + O(\frac{r}{\|y\|} + \frac{r \ln r}{\|x - y\|}))}. \tag{6.43}
$$

Proof Apply the optional stopping theorem to the martingale $\hat{g}(\widehat{S}_{n \wedge \hat{\tau}_{B(y,r)}}, y)$ with the stopping time $\hat{\tau}_{B(y,r)}$ to obtain that

$$
\begin{aligned}
&\frac{a(x) + a(y) - a(x - y)}{a(x)a(y)} \\
&= \hat{g}(x, y) \\
&= \mathbb{P}_x[\hat{\tau}_{B(y,r)} < \infty]\mathbb{E}_x(\hat{g}(\widehat{S}_{\hat{\tau}_{B(y,r)}}, y) \mid \hat{\tau}_{B(y,r)} < \infty)
\end{aligned}
$$

(by (6.41))

$$
= \frac{2a(y) - a(r) + O(r^{-1} + \frac{r}{\|y\|})}{a^2(y)}\mathbb{P}_x[\hat{\tau}_{B(y,r)} < \infty],
$$

and (6.42) follows.

To prove part (ii), we reuse the preceding calculation (with A on the place of $B(y, r)$), but we will need a finer estimate on the term $\mathbb{E}_x(\hat{g}(\widehat{S}_{\hat{\tau}_A}, y) \mid \hat{\tau}_A < \infty)$ in there. Recall (4.37): $\widehat{\text{hm}}_A$ is hm_A biased by a; so, with (3.38) it is straightforward to obtain that

$$\widehat{\text{hm}}_A(z) = \text{hm}_A(z)\left(1 + O\left(\tfrac{r}{\|y\| \ln \|y\|}\right)\right) \tag{6.44}$$

for all $z \in A$. Now, write

$$\mathbb{E}_x(\hat{g}(\widehat{S}_{\hat{\tau}_A}, y) \mid \hat{\tau}_A < \infty)$$
$$= \sum_{z \in A} \hat{g}(z, y)\mathbb{P}_x[\widehat{S}_{\hat{\tau}_A} = z \mid \hat{\tau}_A < \infty]$$

(by (6.41))

$$= \sum_{z \in A}\left(\frac{2}{a(y)} - \frac{a(y - z)}{a^2(y)} + O\left(\tfrac{r}{\|y\| \ln^2 \|y\|}\right)\right)\mathbb{P}_x[\widehat{S}_{\hat{\tau}_A} = z \mid \hat{\tau}_A < \infty]$$
$$= \frac{2}{a(y)} + O\left(\tfrac{r}{\|y\| \ln^2 \|y\|}\right) - \frac{1}{a^2(y)}\sum_{z \in A} a(y - z)\mathbb{P}_x[\widehat{S}_{\hat{\tau}_A} = z \mid \hat{\tau}_A < \infty]$$

(by Theorem 4.16)

$$= \frac{2}{a(y)} + O\left(\tfrac{r}{\|y\| \ln^2 \|y\|}\right) - \frac{1}{a^2(y)}\sum_{z \in A} a(y - z)\widehat{\text{hm}}_A(z)\left(1 + O\left(\tfrac{r}{\|x-y\|}\right)\right)$$

(by (6.44))

$$= \frac{2}{a(y)} + O\left(\tfrac{r}{\|y\| \ln^2 \|y\|}\right)$$
$$- \frac{1}{a^2(y)}\sum_{z \in A} a(y - z)\,\text{hm}_A(z)\left(1 + O\left(\tfrac{r}{\|x-y\|} + \tfrac{r}{\|y\| \ln \|y\|}\right)\right)$$

(by Definition 3.18)

$$= \frac{2}{a(y)} + O\left(\tfrac{r}{\|y\| \ln^2 \|y\|}\right) - \frac{1}{a^2(y)}\left(\text{cap}(A) + O\left(\tfrac{r \ln r}{\|x-y\|} + \tfrac{r \ln r}{\|y\| \ln \|y\|}\right)\right)$$
$$= \frac{2a(y) - \text{cap}(A) + O\left(\tfrac{r \ln r}{\|x-y\|} + \tfrac{r}{\|y\|}\right)}{a^2(y)},$$

and we then obtain (6.43) in an analogous way.   □

6.3.3 Harmonic measure and capacities

Before proceeding, recall the following immediate consequence of (4.31): for any finite $A \subset \mathbb{Z}^2$ such that $0 \in A$, we have

$$\text{cap}(A) = \lim_{\|x\| \to \infty} a(x) \mathbb{P}_x[\hat{\tau}_A < \infty]. \tag{6.45}$$

Next, we need estimates for the $\widehat{S}$-capacity of a "distant" set[15] and, in particular, of a ball which does not contain the origin. This is a cleaner version of lemma 3.9 of [26].

Lemma 6.15. *Assume that* $\|y\| > 2r \geq 1$.

(i) *We have*

$$\text{cap}(\{0\} \cup B(y, r)) = \frac{a^2(y)}{2a(y) - a(r) + O(r^{-1} + \frac{r}{\|y\|})}. \tag{6.46}$$

(ii) *For a finite set A such that $y \in A$, denote $r = 1 + \max_{z \in A} \|y - z\|$ and assume that $\|y\| > 2r$. Then*

$$\text{cap}(\{0\} \cup A) = \frac{a^2(y)}{2a(y) - \text{cap}(A) + O(\frac{r}{\|y\|})}. \tag{6.47}$$

Proof This immediately follows from Lemma 6.14 and (6.45) (observe that $a(x) - a(x - y) \to 0$ as $x \to \infty$). $\square$

We also need to compare the harmonic measure on a set (again, distant from the origin) to the entrance measure of the $\widehat{S}$-walk started far away from that set (this would be a partial analogue of Theorem 3.20 for the conditioned walk).

Lemma 6.16. *Assume that A is a finite subset of $\mathbb{Z}^2$ and $y_0 \in A$ and $x \notin A$ are such that $\|y_0 - x\| \leq \|y_0\|/2$ and $\text{dist}(x, A) \geq 18 \, \text{diam}(A) + 1$. Assume additionally that (finite or infinite) $A' \subset \mathbb{Z}^2$ is such that $\text{dist}(A, A') \geq \text{dist}(x, A) + 1$. Then it holds that*

$$\mathbb{P}_x[\widehat{S}_{\hat{\tau}_A} = y \mid \hat{\tau}_A < \hat{\tau}_{A'}] = \widehat{\text{hm}}_A(y)(1 + O(\frac{\Psi \, \text{diam}(A)}{\text{dist}(x, A)})), \tag{6.48}$$

where $\Psi = (a(\text{dist}(x, A)) - \text{cap}(A)) \vee 1$.

Proof The proof essentially mimics that of Theorem 3.20, so we give only a sketch. As before, with $r = \text{diam}(A)$ and $s := \|x\|$, it is enough to

---

[15] I.e., for the "usual" capacity of the union of that set and $\{0\}$.

prove (6.48) for $A' = B(y_0, s)^C$. Again, since on its way to A the walker has to pass through $\partial B(y_0, \frac{2}{3}s)$, it would then be enough to prove that

$$P_z[\widehat{S}_{\hat{\tau}_A} = y \mid \hat{\tau}_A < \hat{\tau}_{B(y_0,s)^C}] = \widehat{hm}_A(y)(1 + O(\tfrac{\Psi r}{s})) \tag{6.49}$$

for any $z \in \partial B(y_0, \frac{2}{3}s)$. With the help of Lemma 4.5 and (3.64), we obtain that $P_z[\hat{\tau}_A < \hat{\tau}_{B(y_0,s)^C}] \geq O(1/\Psi)$. So, we write

$$P_z[\widehat{S}_{\hat{\tau}_A} = y, \hat{\tau}_A < \hat{\tau}_{B(y_0,s)^C}]$$

$$= P_z[\widehat{S}_{\hat{\tau}_A} = y] - P_z[\widehat{S}_{\hat{\tau}_A} = y, \hat{\tau}_{B(y_0,s)^C} < \hat{\tau}_A]$$

$$= P_z[\widehat{S}_{\hat{\tau}_A} = y] - \sum_{z' \in B(y_0,s)^C} P_z[\widehat{S}_{\hat{\tau}_{B(y_0,s)^C}} = z', \hat{\tau}_{B(y_0,s)^C} < \hat{\tau}_A] P_{z'}[\widehat{S}_{\hat{\tau}_A} = y]$$

(by Theorem 4.16)

$$= \widehat{hm}_A(y)(1 + O(\tfrac{r}{s})) - P_z[\hat{\tau}_{B(y_0,s)^C} < \hat{\tau}_A] \widehat{hm}_A(y)(1 + O(\tfrac{r}{s}))$$

$$= \widehat{hm}_A(y)(P_z[\hat{\tau}_A < \hat{\tau}_{B(y_0,s)^C}] + O(\tfrac{r}{s}))$$

and then conclude the proof by dividing the last expression by $P_z[\hat{\tau}_A < \hat{\tau}_{B(y_0,s)^C}]$. $\qquad\qquad\square$

6.3.4 Proofs for random interlacements

First of all, we finish the proof of Theorem 6.8.

Proof of Theorem 6.8, parts (iii) through (v). Let us recall the fundamental formula (6.24) for the random interlacements and the relation (3.36). Then, the statement (iv) follows from (3.86) and from (6.29) (see also Exercise 6.7), while (v) is a consequence of Lemma 6.15 (i).

Finally, observe that, by symmetry (which gives us that $\text{cap}(A \cup \{x\}) = \text{cap}(\{0\} \cup A')$ with $A' = A - x$, a translate of A), Theorem 6.8 (ii), and Lemma 6.15 (ii), we have

$$P[A \subset \mathcal{V}^\alpha \mid x \in \mathcal{V}^\alpha]$$

$$= \exp\left(-\pi\alpha(\text{cap}(A \cup \{x\}) - \text{cap}(\{0, x\}))\right)$$

$$= \exp\left(-\pi\alpha\left(\frac{a^2(x)}{2a(x) - \text{cap}(A) + O(\frac{r}{\|x\|})} - \frac{a(x)}{2}\right)\right)$$

$$= \exp\left(-\pi\alpha\left(\frac{a^2(x) - a^2(x) + \frac{1}{2}a(x)\,\text{cap}(A) + O(\frac{r\ln\|x\|}{\|x\|})}{2a(x) - \text{cap}(A) + O(\frac{r}{\|x\|})}\right)\right)$$

$$= \exp\left(-\frac{\pi\alpha}{4}\operatorname{cap}(A)\frac{1+O(\frac{r}{\|x\|\ln r})}{1-\frac{\operatorname{cap}(A)}{2a(x)}+O(\frac{r}{\|x\|\ln x})}\right),$$

thus proving part (iii). □

Proof of Theorem 6.10 (iii) We start by observing that the first part of (iii) follows from the bound (6.33) and Borel–Cantelli. So, let us concentrate on proving (6.33). Recall the following elementary fact: let N be a Poisson random variable with parameter λ, and $Y_1, Y_2, Y_3, \ldots$ be independent (also of N) random variables with exponential distribution with parameter p (that is, mean value p^{-1}). Let $\Theta = \sum_{j=1}^{N} Y_j$ be the corresponding compound Poisson random variable, with $\mathbb{E}\Theta = \lambda p^{-1}$. Its moment generating function $M_\Theta(h) = \exp\left(\frac{\lambda h}{p-h}\right)$ is easily computed, and, after some straightforward calculations, we then can write Chernoff's bound for it: for all $b > 1$

$$\mathbb{P}[\Theta \geq b\lambda p^{-1}] \leq \exp\left(-\lambda(\sqrt{b}-1)^2\right). \tag{6.50}$$

Assume that $\alpha < 1$. Fix $\beta \in (0,1)$, which will be later taken close to 1, and fix some set of non-intersecting disks $B'_1 = B(x_1, r^\beta), \ldots, B'_{k_r} = B(x_{k_r}, r^\beta) \subset B(r) \setminus B(r/2)$, with cardinality $k_r = \frac{1}{4}r^{2(1-\beta)}$. Denote also $B_j := B(x_j, \frac{r^\beta}{\ln^3 r^\beta})$, $j = 1, \ldots, k_r$.

Before going to the heart of the matter, we briefly sketch the strategy of proof (also, one may find it helpful to look at Figure 6.8). We start by showing that at least a half of these disks B_j will receive at most $b\frac{2\alpha\ln^2 r}{3\ln\ln r^\beta}$ excursions from ∂B_j to $\partial B'_j$, where $b > 1$ is a parameter (in fact, the aforementioned number of excursions is larger by factor b than the typical number of excursions on level α). Moreover, using the method of soft local times of Section 5.1, we couple such excursions from $\mathrm{RI}(\alpha)$ with a slightly larger number of independent $\widehat{S}$-excursions: with overwhelming probability, the trace on $\bigcup_j B_j$ of the latter excursion process contains the trace of the former, so the vacant set $\mathcal{V}^\alpha$ restricted to disks B_j is smaller than the set of unvisited points by the independent process. By independence, it will be possible to estimate the probability of leaving that many balls partially uncovered, and this will conclude the proof.

Let us observe that the number of $\widehat{S}$-walks in $\mathrm{RI}(\alpha)$ intersecting a given disk B_j has the Poisson law with parameter $\lambda = (1 + o(1))\frac{2\alpha}{2-\beta}\ln r$. Indeed, the law is Poisson by construction, the parameter $\pi\alpha \operatorname{cap}(B_j \cup \{0\})$ is found in (6.24) and then estimated using Lemma 6.15 (i).

Next, by Lemma 6.14 (i), the probability that the walk $\widehat{S}$ started from any $y \in \partial B'_j$ does not hit B_j is $(1 + o(1))\frac{3\ln\ln r^\beta}{(2-\beta)\ln r}$. Therefore, after the first

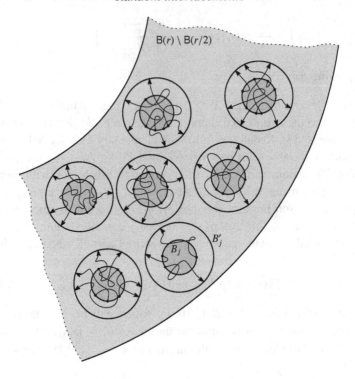

Figure 6.8 On the proof of Theorem 6.10 (iii). With high probability, at least a positive proportion of the inner circles is not completely covered.

visit to $\partial B'_j$, each $\widehat{S}$-walk generates a number of excursions between ∂B_j and $\partial B'_j$, which is dominated by a geometric law (supported on $\{0, 1, 2, \ldots\}$) with success parameter $p' = (1 + o(1))\frac{3\ln\ln r^\beta}{(2-\beta)\ln r}$. Recall also that the integer part of an exponential(u) random variable has a geometric[16] distribution with success probability $(1 - e^{-u})$. So, with $p = -\ln(1 - p')$, the total number $\widehat{E}_\alpha^{(j)}$ of excursions between ∂B_j and $\partial B'_j$ in RI(α) can be dominated by a compound Poisson law with exponential(p) terms in the sum with expectation

$$\lambda p^{-1} = (1 + o(1))\frac{2\alpha \ln^2 r}{3\ln\ln r^\beta}.$$

---

[16] Again, the one supported on $\{0, 1, 2, \ldots\}$.

Then, using (6.50), we obtain for $b > 1$

$$\mathbb{P}\left[\widehat{E}_\alpha^{(j)} \geq b\frac{2\alpha \ln^2 r}{3 \ln \ln r^\beta}\right] \leq \exp\left(-(1 + o(1))(\sqrt{b} - 1)^2 \frac{2\alpha}{2 - \beta} \ln r\right)$$

$$= r^{-(1+o(1))(\sqrt{b}-1)^2 \frac{2\alpha}{2-\beta}}. \tag{6.51}$$

Now, let W_b be the set

$$W_b = \left\{j \leq k_r : \widehat{E}_\alpha^{(j)} < b\frac{2\alpha \ln^2 r}{3 \ln \ln r^\beta}\right\}.$$

Combining (6.51) with Markov inequality, we obtain

$$\frac{k_r}{2}\mathbb{P}[|\{1, \ldots, k_r\} \setminus W_b| > k_r/2] \leq \mathbb{E}\big|\{1, \ldots, k_r\} \setminus W_b\big|$$

$$\leq k_r r^{-(1+o(1))(\sqrt{b}-1)^2 \frac{2\alpha}{2-\beta}},$$

so

$$\mathbb{P}[|W_b| \geq k_r/2] \geq 1 - 2r^{-(1+o(1))(\sqrt{b}-1)^2 \frac{2\alpha}{2-\beta}}. \tag{6.52}$$

Assume that $1 < b < \alpha^{-1}$ and $\beta \in (0, 1)$ is close enough to 1, so that $\frac{b\alpha}{\beta^2} < 1$. We denote by $\widehat{\mathcal{E}}_1^{(j)}, \ldots, \widehat{\mathcal{E}}_{\widehat{E}_\alpha^{(j)}}^{(j)}$ the excursions of $\mathrm{RI}(\alpha)$ between ∂B_j and $\partial B'_j$. Also, let $\widetilde{\mathcal{E}}_1^{(j)}, \widetilde{\mathcal{E}}_2^{(j)}, \widetilde{\mathcal{E}}_3^{(j)}, \ldots$ be a sequence of i.i.d. $\widehat{S}$-excursions between ∂B_j and $\partial B'_j$, started with the law $\widehat{\mathrm{hm}}_{B_j}$; these sequences are also assumed to be independent from each other.

Abbreviate $m = b\frac{2\alpha \ln^2 r}{3 \ln \ln r^\beta}$. Next, for $j = 1, \ldots, k_r$ we consider the events

$$D_j = \left\{\{\widehat{\mathcal{E}}_1^{(j)}, \ldots, \widehat{\mathcal{E}}_{\widehat{E}_\alpha^{(j)}}^{(j)}\} \subset \{\widetilde{\mathcal{E}}_1^{(j)}, \ldots, \widetilde{\mathcal{E}}_{(1+\delta)m}^{(j)}\}\right\}.$$

Lemma 6.17. *One can construct the excursions* $(\widehat{\mathcal{E}}_k^{(j)}, k = 1, \ldots, \widehat{E}_\alpha^{(j)})$, $(\widetilde{\mathcal{E}}_k^{(j)}, k = 1, 2, 3, \ldots)$, $j = 1, \ldots, k_r$, *on a same probability space in such a way that for a fixed $C' > 0$*

$$\mathbb{P}[D_j^C] \leq \exp\left(-C'\frac{\ln^2 r}{(\ln \ln r)^2}\right), \tag{6.53}$$

for all $j = 1, \ldots, k_r$.

Proof The inequality (6.53) follows from Lemma 6.12 (observe that, by Lemma 6.16, the parameter ν in Lemma 6.12 can be anything exceeding $O(1/\ln^2 r)$, so we choose e.g. $\nu = (\ln \ln r)^{-1}$). $\square$

We continue the proof of part (iii) of Theorem 6.10. Define

$$D = \bigcap_{j \leq k_r} D_j;$$

using (6.53), by the union bound we obtain the subpolynomial estimate

$$\mathbb{P}[D^C] \leq \frac{1}{4} r^{2(1-\beta)} \exp\left(-C' \frac{\ln^2 r}{(\ln \ln r)^2}\right). \tag{6.54}$$

Let $\delta > 0$ be such that $(1 + \delta)\frac{b\alpha}{\beta^2} < 1$. Define the events

$$M_j = \left\{B_j \text{ is completely covered by } \widetilde{\mathcal{E}x}_1^{(j)} \cup \cdots \cup \widetilde{\mathcal{E}x}_{(1+\delta)m}^{(j)}\right\}.$$

Then for all $j \leq k_r$, it holds that

$$\mathbb{P}[M_j] \leq \frac{1}{5} \tag{6.55}$$

for all large enough r. Indeed, if the $\widetilde{\mathcal{E}x}$'s were independent SRW excursions, the preceding inequality (with any fixed constant in the right-hand side) would be just a consequence of Lemma 6.13. On the other hand, Lemma 4.5 implies that the first $(1+\delta)m\,\widehat{S}$-excursions can be coupled with SRW excursions with high probability, so (6.55) holds for $\widehat{S}$-excursions as well.

Next, define the set

$$\widetilde{W} = \{j \leq k_r : M_j^C \text{ occurs}\}.$$

Since the events $(M_j, j \leq k_r)$ are independent, by (6.55) we have (recall that k_r is of order $r^{2(1-\beta)}$)

$$\mathbb{P}\left[|\widetilde{W}| \geq \frac{3}{5} k_r\right] \geq 1 - \exp\left(-Cr^{2(1-\beta)}\right) \tag{6.56}$$

for all r large enough.

Observe that, by construction, on the event D we have $\mathcal{V}^\alpha \cap B'_j \neq \emptyset$ for all $j \in \widetilde{W} \cap W_b$. So, using (6.52), (6.54), and (6.56), we obtain

$$\mathbb{P}\left[\mathcal{V}^\alpha \cap (B(r) \setminus B(r/2)) = \emptyset\right]$$

$$\leq \mathbb{P}\left[|W_b| < \frac{k_r}{2}\right] + \mathbb{P}[D^C] + \mathbb{P}\left[|\widetilde{W}| < \frac{3k_r}{5}\right]$$

$$\leq 2r^{-(1+o(1))(\sqrt{b}-1)^2 \frac{2\alpha}{2-\beta}} + \exp\left(-Cr^{2(1-\beta)}\right)$$

$$+ \frac{1}{4} r^{2(1-\beta)} \exp\left(-C' \frac{\ln^2 r}{(\ln \ln r)^2}\right).$$

Since $b \in (1, \alpha^{-1})$ can be arbitrarily close to α^{-1} and $\beta \in (0, 1)$ can be arbitrarily close to 1, this concludes the proof of (6.33).   □

Proof of Theorem 6.10 (ii)   To complete the proofs of the results of Section 6.2, it remains to show that $|\mathcal{V}^\alpha| < \infty$ a.s. for $\alpha > 1$. First, we establish the following: if $r \geq 2$, then for any $x \in \partial B(2r)$ and $y \in B(r) \setminus B(r/2)$, it holds that

$$P_x[\hat{\tau}_y < \hat{\tau}_{B(r \ln r)}] = \frac{\ln \ln r}{\ln r}(1 + o(1)). \qquad (6.57)$$

Indeed, Lemma 4.20 implies that

$$P_x[\hat{\tau}_y < \hat{\tau}_{B(r \ln r)}]$$
$$= \frac{a(r \ln r)(a(x) + a(y) - a(x - y)) - a(x)a(y)(1 + O(\frac{1}{\ln^2 r}))}{a(x)(2a(r \ln r) - a(y)(1 + O(\frac{1}{\ln^2 r})))}.$$

We then use (3.36) to obtain (after dividing by $\frac{4}{\pi^2}$) that the numerator in the preceding expression is $\ln^2 r + \ln r \ln \ln r + O(\ln r) - (\ln^2 r + O(\ln r)) = \ln r \ln \ln r + O(\ln r)$, while the denominator is $\ln^2 r + O(\ln r \ln \ln r)$; this clearly implies (6.57).

Now the goal is to prove that, for $\alpha > 1$

$$\mathbb{P}[\exists y \in B(r) \setminus B(r/2) \text{ such that } y \in \mathcal{V}^\alpha] \leq r^{-\frac{\alpha}{2}(1 - \alpha^{-1})^2(1 + o(1))}. \qquad (6.58)$$

This would imply that the set $\mathcal{V}^\alpha$ is a.s. finite, since

$$\{|\mathcal{V}^\alpha| = \infty\} = \{\mathcal{V}^\alpha \cap (B(2^n) \setminus B(2^{n-1})) \neq \emptyset \text{ for infinitely many } n\}, \qquad (6.59)$$

and the Borel–Cantelli lemma together with (6.58) imply that the probability of the latter event equals 0.

Let $E_{\alpha,r}$ be the number of $\widehat{S}$-excursions of RI(α) between $\partial B(r)$ and $\partial B(r \ln r)$. Analogously to (6.51) (but using Corollary 4.7 instead of Lemma 6.14 (i)), it is straightforward to show that, for $b < 1$,

$$\mathbb{P}\left[E_{\alpha,r} \leq b\frac{2\alpha \ln^2 r}{\ln \ln r}\right] \leq r^{-2\alpha(1 - \sqrt{b})^2(1 + o(1))}. \qquad (6.60)$$

Now (6.57) implies that for $y \in B(r) \setminus B(r/2)$

$$\mathbb{P}\left[y \text{ is uncovered by first } b\frac{2\alpha \ln^2 r}{\ln \ln r} \text{ excursions}\right]$$
$$\leq \left(1 - \frac{\ln \ln r}{\ln r}(1 + o(1))\right)^{b\frac{2\alpha \ln^2 r}{\ln \ln r}}$$
$$= r^{-2b\alpha(1 + o(1))}, \qquad (6.61)$$

so, using the union bound,

$$\mathbb{P}\left[\exists y \in B(r) \setminus B(r/2) : y \in \mathcal{V}^\alpha, E_{\alpha,r} > b\frac{2\alpha \ln^2 r}{\ln \ln r}\right] \le r^{-2(b\alpha-1)(1+o(1))}. \quad (6.62)$$

Using (6.60) and (6.62) with $b = \frac{1}{4}(1 + \frac{1}{\alpha})^2$ (indeed, with b as before we have $b\alpha - 1 = \frac{\alpha}{4}(1 + \frac{2}{\alpha} + \frac{1}{\alpha^2}) - 1 = \frac{\alpha}{4}(1 - \frac{1}{\alpha})^2 > 0$), we conclude the proof of (6.58) and of Theorem 6.10 (ii).   □

6.4 Exercises

Exercise 6.1. Given $x \ne y$, find $\varepsilon = \varepsilon(x, y)$ such that for all $u > 0$ it holds that

$$\mathbb{P}[\{x, y\} \subset \mathcal{V}^u] = \mathbb{P}[x \in \mathcal{V}^{(1-\varepsilon)u}]\mathbb{P}[y \in \mathcal{V}^{(1-\varepsilon)u}]. \quad (6.63)$$

What is the asymptotic behaviour of $\varepsilon(x, y)$ as $\|x - y\| \to \infty$?

Exercise 6.2. For $d \ge 2$, let $z \in \partial B(r)$, $r \ge 1$; also, fix $h \in [1, r]$.

(i) Assume that $x \in \mathbb{Z}^d \setminus B(r)$ is such that $\|x - z\| \ge h$. Prove that there exists a positive constant c such that

$$\mathbb{P}_x[S_{\tau_A} = x] \le ch^{-(d-1)}. \quad (6.64)$$

(ii) Next, assume that $x \in \mathbb{Z}^d \setminus B(r)$ is such that $\|x - z\| \le 2h$ and also $\text{dist}(x, B(r)) \ge h$. Prove that, for some $c' > 0$

$$\mathbb{P}_x[S_{\tau_A} = x] \ge c'h^{-(d-1)}. \quad (6.65)$$

Exercise 6.3. Show (formally) that Theorem 6.4 implies Theorem 6.3, and Theorem 6.3 implies Theorem 6.2. Can Theorem 6.4 be also used to deduce a decoupling result for the field of *local times* of random interlacements?

Exercise 6.4. Show that using the soft local times method with a *marked* Poisson point process on ∂A (recall Figure 6.4) is indeed an equivalent way of constructing random interlacements' excursions.

Exercise 6.5. For two-dimensional random interlacements, prove that the vacant set $\mathcal{V}^\alpha$ does not percolate for any $\alpha > 0$.

Exercise 6.6. Similarly to Remark 6.9 (b), one can also estimate the "local rate" away from the origin. Consider the situation depicted in Figure 6.9 (where s is large), and assume that $\text{cap}(A_2) \ll \ln s$. Show that (6.30) then reveals a "local rate" equal to $\frac{2}{7}\alpha$, that is, $\mathbb{P}[A_2 \subset \mathcal{V}^\alpha \mid x \in \mathcal{V}^\alpha] = \exp(-\frac{2}{7}\pi\alpha \, \text{cap}(\{0\} \cup A_2)(1 + o(1)))$.

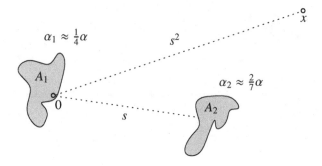

Figure 6.9 What the "local rate" looks like if we condition on the event that a "distant" site is vacant.

Exercise 6.7. Do the calculations in the proof of Theorem 6.8 (iv) (those that involve using (3.86)).

Exercise 6.8. For any $\varepsilon > 0$, prove that $\mathbb{P}[|\mathcal{V}^\alpha| < \infty] = 1$ implies $\mathbb{P}[\mathcal{V}^{\alpha+\varepsilon} = \{0\}] > 0$ without invoking the Harris–FKG inequality (this therefore provides another proof that $\mathbb{P}[\mathcal{V}^\alpha = \{0\}] > 0$ for any $\alpha > 1$).

Exercise 6.9. Can you prove the result of Lemma 6.13 in a more direct way (i.e., without using known facts for the cover time of the torus by SRW)?

Exercise 6.10. Prove the claim of Remark 4.19 by proving that (analogously to the proof of (6.33)), with *any* fixed $\delta > 0$,

$$\mathbb{P}[D_n \subset \widehat{S}_{[0,\infty)}] \leq n^{-2+\delta}$$

for all large enough n; the claim then would follow from the (first) Borel–Cantelli lemma.

Exercise 6.11. Write a more detailed proof of (6.60).

Exercise 6.12. If it makes sense considering random interlacements in two dimensions, then can we probably do so in dimension 1 as well? Think about how this can be done.

Hints and solutions to selected exercises

Exercise 2.2.

Unfortunately, no. If S_n is a d-dimensional SRW, then $\mathbb{P}_0[S_2 = 0] = (\frac{1}{2d} \times \frac{1}{2d}) \times (2d) = \frac{1}{2d}$. If we want the claim in the exercise to be true, this would then mean that $\frac{1}{2d} = (\frac{1}{2})^d$ or $d = 2^{d-1}$, which holds only for $d = 1, 2$.

Exercise 2.3.

You may find it useful to read [35].

Exercise 2.6.

See the proof of proposition 6.1 in [80].

Exercise 2.7.

Set $\pi(0) = 1$, and then prove by induction that

$$\pi(x) = \frac{q_0 \cdots q_{x-1}}{p_1 \cdots p_x}$$

for $x > 0$ (and analogously for $x < 0$).

Exercise 2.8.

Use the cycle criterion (Theorem 2.2) with e.g. the cycle $(0,0) \to (0,1) \to (1,1) \to (1,0) \to (0,0)$.

Exercise 2.9.

Hint: define a scalar product $\langle f, g \rangle = \sum_{x \in \Sigma} \pi(x) f(x) g(x)$ and show that P is self-adjoint.

Exercise 2.12.

We can rewrite (2.20) as

$$\frac{1}{\frac{1}{a+c} + \frac{1}{b+d}} \geq \frac{1}{\frac{1}{a} + \frac{1}{b}} + \frac{1}{\frac{1}{c} + \frac{1}{d}}.$$

185

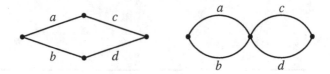

Figure H.1 For Exercise 2.12: think of $a, b, c, d > 0$ as resistances, and note that the effective resistance (from left to right) of the left circuit is greater than or equal to the effective resistance of the right circuit.

Then, just look at Figure H.1.

Exercise 2.15.

Fix an arbitrary $x_0 \in \Sigma$, set $A = \{x_0\}$, and

$$f(x) = \mathbb{P}_x[\tau_{x_0} < \infty] \text{ for } x \in \Sigma$$

(so, in particular, $f(x_0) = 1$). Then (2.12) holds with equality for all $x \neq x_0$, and, by transience, one can find $y \in \Sigma$ such that $f(y) < 1 = f(x_0)$.

Exercise 2.16.

Let $p = p(n, n+1)$ (for all n), and assume for definiteness that $p > \frac{1}{2}$. Consider the function $f(x) = (\frac{1-p}{p})^x$ and the set $A = (-\infty, 0]$; then use Theorem 2.5.

Note also that, for proving that this random walk is transient, one may also use theorem 2.5.15 of [71] (which we did not consider in this book) together with a simpler function $f(x) = x$. There are many different Lyapunov function tools that one may use!

Exercise 2.17.

Hint: use the Lyapunov function $f(x) = (1 - \delta)^x$ for small enough $\delta > 0$.

Exercise 2.18.

Quite analogously to (2.15) through (2.17), it is elementary to obtain for $f(x) = \|x\|^{-\alpha}$

$$\mathbb{E}(f(X_{n+1}) - f(X_n) \mid X_n = x)$$
$$= -\alpha \|x\|^{-\alpha-2} \left(\frac{1}{2} - \left(1 + \frac{\alpha}{2} \right) \frac{1}{d} + O(\|x\|^{-1}) \right).$$

The inequality $\frac{1}{2} - (1 + \frac{\alpha}{2}) \frac{1}{d} > 0$ solves to $\alpha < d - 2$, so any fixed $\alpha \in (0, d-2)$ will do the job.

By the way, in your opinion, is it surprising that the "critical" value for α is equal to $d - 2$? (To answer this question, it is maybe a good idea to read Section 3.1.)

Exercise 2.20.

Hint: being X_n the two-dimensional walk, define first its *covariance matrix* by $M := \mathbb{E}_0((X_1)^\top X_1)$. Find a suitable linear transformation[1] of the process for which M will become the identity matrix. Then use the same Lyapunov function that worked for the simple random walk.

Exercise 2.23 (i).

Fix an arbitrary $x_0 \in \Sigma$, and set $A = \{x_0\}$. Observe that, for $x \neq x_0$,

$$\mathbb{E}_x \tau_{x_0} = \sum_{y \in \Sigma} p(x, y) \mathbb{E}_y (1 + \tau_{x_0}),$$

and that

$$\sum_{y \in \Sigma} p(x_0, y) \mathbb{E}_y \tau_{x_0} = \mathbb{E}_{x_0} \tau_{x_0}^+ < \infty,$$

so the function $f(x) = \mathbb{E}_x \tau_{x_0}$ satisfies (2.21) and (2.22) with $\varepsilon = 1$.

Exercise 2.24 (ii).

Hint: apply theorem 2.6.2 of [71] to the process $X_n = \ln(Y_n + 2)$.

Exercise 2.25.

Hint: First, show that it is enough to prove that there exists large enough R such that, regardless of the starting position, the process visits $B(R)$ a.s.. Then, similarly to the proof of the recurrence of two-dimensional SRW, use Theorem 2.4 with a suitable Lyapunov function (with the advantage that here the process is *really* radially symmetric). In particular, will $f(x) = \ln \|x\|$ work here?

Exercise 2.26.

Note that the calculation (2.15) is dimension independent, and (2.16) remains valid as well, with obvious changes. Then obtaining (2.23) is straightforward (use (2.15) with $\alpha = 1$ and observe that the factor $\frac{1}{4}$ in the next display after (2.15) will become $\frac{1}{2d}$ in the general case). As for (2.24), show first that

$$\mathbb{E}_x(\|S_1\|^2 - \|x\|^2) = 1 \text{ for all } x \in \mathbb{Z}^d,$$

[1] Why does it exist?

and then use (2.23) together with the identity $(b - a)^2 = b^2 - a^2 - 2a(b - a)$ with $a = \|x\|, b = \|S_1\|$.

Exercise 2.27.

Hint: try using the following Lyapunov functions: $f(x) = x^2$ for (i), $f(x) = x^\alpha$ for some $\alpha > 0$ for (ii), and $f(x) = x^{-\alpha}$ for (iii). Note that α will depend on ε in (ii) and (iii)!

Exercise 2.28.

Hint: use Exercise 2.26.

Exercise 3.3.

Hint: first, prove that the (multivariate) characteristic function of S_n (starting at the origin) equals $\Phi^n(\theta)$. Then, think how to *extract* $\mathbb{P}_0[S_n = x]$ from there.

Exercise 3.5.

See exercise 6.18 of [63].

Exercise 3.6.

Hint: use Exercise 2.28.

Exercise 3.7.

Hint: show that $Y_n = G(S_n - y) + N_y^{(n-1)}$ is a martingale (recall Exercise 3.2), then use the optional stopping theorem.

Exercise 3.8.

Hint: first, use a reversibility argument to relate the probability in (3.73) to restricted Green's functions. To deal with these Green's functions, first prove that the walk *escapes* from y to $\partial B((\frac{1}{4} + \delta)r)$ with probability of order r^{-1}; this can be done analogously to the argument right after (3.17) (or with Lyapunov functions). Then use (3.72) (or Theorem 3.13 in two dimensions).

Exercise 3.9.

For $x \notin A$, using the definition of G and the strong Markov property, let us write,

$$G(x, A) = \sum_{y \in A} \mathbb{P}_x[\tau_A < \infty, S_{\tau_A} = y] G(y, A)$$

$$\geq \left(\sum_{y \in A} \mathbb{P}_x[\tau_A < \infty, S_{\tau_A} = y] \right) \times \min_{y \in A} G(y, A)$$

$$= \mathbb{P}_x[\tau_A < \infty] \min_{y \in A} G(y, A),$$

and, analogously,

$$G(x, A) \leq \mathbb{P}_x[\tau_A < \infty] \max_{y \in A} G(y, A),$$

so

$$\frac{G(x, A)}{\max_{y \in A} G(y, A)} \leq \mathbb{P}_x[\tau_A < \infty] \leq \frac{G(x, A)}{\min_{y \in A} G(y, A)}. \tag{H.1}$$

From the asymptotic expression (3.5) for Green's function, it is straightforward to obtain that

$$\lim_{x \to \infty} \frac{\|x\|}{\gamma_d} G(x, A) = |A|,$$

so multiplying (H.1) by $\frac{\|x\|}{\gamma_d}$ and using Proposition 3.4, we obtain the desired result.

Exercise 3.10.

First of all, note that $\mathrm{Es}_A \in \mathcal{K}_A$ by (3.11), so it is enough to prove that $\sum_{x \in A} h(x) \leq \mathrm{cap}(A)$ for any $h \in \mathcal{K}_A$. Let us abbreviate $v_A(x) = \mathbb{P}_x[\tau_A < \infty]$, so that (in matrix notation) (3.11) becomes $v_A = G \, \mathrm{Es}_A$. Now, for two (nonnegative) functions $f, g : \mathbb{Z}^d \to \mathbb{R}$, define the usual scalar product by $(f, g) = \sum_{x \in \mathbb{Z}^d} f(x) g(x)$. We argue that, by symmetry, G is a self-adjoint linear operator in the sense that $(Gf, g) = (f, Gg)$. Then, for any $h \in \mathcal{K}_A$, write ($\mathbf{1}$ being the function with value 1 at all x)

$$\sum_{x \in A} h(x) = (h, v_A) = (h, G \, \mathrm{Es}_A) = (Gh, \mathrm{Es}_A) \leq (\mathbf{1}, \mathrm{Es}_A) = \mathrm{cap}(A),$$

and we are done.

Exercise 3.13.

Use the second inequality in (3.14) to obtain that, when the series in (3.19) converges, it holds that

$$\sum_{k=1}^{\infty} \mathbb{P}_0[\tau_{A_k} < \infty] < \infty;$$

then use Borel–Cantelli.

Exercise 3.14.

Analogously to the previous exercise, it is straightforward to obtain that, in this case,

$$\sum_{k=1}^{\infty} \mathbb{P}_0[\tau_{A_k} < \infty] = \infty,$$

but obtaining the recurrence of A from this is not immediate (note that the events $\{\tau_{A_k} < \infty\}$, $k \geq 1$, need not be independent). One possible workaround is the following:

(i) Divide the sum in (3.19) into four sums:

$$\sum_{j=1}^{\infty} \frac{\text{cap}(A_{4j+i})}{2^{(d-2)(4j+i)}}, \quad i = 0, 1, 2, 3.$$

Note that at least one of these sums must diverge. For definiteness, assume that it is the first one (i.e., with $i = 0$).

(ii) For $j \geq 2$, consider any $y \in \partial B(2^{4j-2})$, and, using the first inequality in (3.14) together with (3.5), show that $\mathbb{P}_y[\tau_{A_{4j}}] \geq c_1 2^{-(d-2)(4j)} \text{cap}(A_{4j})$.

(iii) Next, for any $z \in \partial B(2^{4j+2})$ show that

$$\mathbb{P}_y[\tau_{A_{4j}}] \leq c_2 \frac{\text{cap}(A_{4j})}{2^{4j(d-2)}},$$

where $c_2 < c_1$. Note that A_{4j} is *much* closer to $\partial B(2^{4j-2})$ than to $\partial B(2^{4j+2})$ – consider drawing a picture to see it better!

(iv) Using (ii) and (iii), obtain that, for any $y \in \partial B(2^{4j-2})$,

$$\mathbb{P}_y[\tau_{A_{4j}} < \tau_{\partial B(2^{4j+2})}] \geq c_3 \frac{\text{cap}(A_{4j})}{2^{4j(d-2)}},$$

i.e., on its way from $\partial B(2^{4j-2})$ to $\partial B(2^{4j+2})$, the walker will hit A_{4j} with probability at least of order $2^{-4j(d-2)} \text{cap}(A_{4j})$.

(v) Argue that the "if" part of Theorem 3.7 follows from (iv) in a straightforward way.

Exercise 3.16.

Of course, it is probably possible to construct many such examples; one possibility (in three dimensions) is to consider a set of the form $\{b_k e_1, k \geq 1\}$, where (b_k) is a strictly increasing sequence of positive integer numbers. Then, it is clear that the expected number of visits to this set is infinite if and only if $\sum_{k=1}^{\infty} b_k^{-1} = \infty$. As shown in the paper [14], it is possible to construct a transient set for which the aforementioned series sums to infinity.

Exercise 3.20.

Proof 1: note that (3.75) is equivalent to

$$\text{cap}(A) = \frac{|A|}{\sum_{y \in A} \text{hm}_A(y) G(y, A)};$$

proceed as in the solution of Exercise 3.9[2] and then use basic properties of the harmonic measure.

Proof 2: note that (3.75) is equivalent to

$$\sum_{y \in A} \text{Es}_A(y) G(y, A) = |A|,$$

and then obtain it directly from (3.11) by summing in $y \in A$.

Exercise 3.21.

Hint: recall the calculation after (3.17).

Exercises 3.22 and 3.23.

See the beginning of Section 6.1.

Exercise 3.24.

Let H be a finite subset of $\mathbb{Z}^d$ such that $(A \cup B) \subset H$. We have

$$\mathbb{P}_{\text{hm}_H}[\tau_B < \infty] \geq \mathbb{P}_{\text{hm}_H}[\tau_A < \infty, \exists m \geq \tau_A \text{ such that } S_m \in B]$$

$$\text{(by (3.77))}$$

$$= \mathbb{P}_{\text{hm}_H}[\tau_A < \infty] \mathbb{P}_{\text{hm}_A}[\tau_B < \infty].$$

On the other hand, (3.76) implies that $\mathbb{P}_{\text{hm}_H}[\tau_A < \infty] = \frac{\text{cap}(A)}{\text{cap}(H)}$ and $\mathbb{P}_{\text{hm}_H}[\tau_B < \infty] = \frac{\text{cap}(B)}{\text{cap}(H)}$, so, inserting this to the preceding calculation, we obtain (3.78).

Exercise 3.25.

First, recall (3.24) – we now need to do a finer analysis of it. Let us rewrite (3.24) in the following way:

$$\mathbb{P}_x[S_{\tau_A} = y \mid \tau_A < \infty]$$
$$= \frac{G(y, x) \text{Es}_A(y)}{\mathbb{P}_x[\tau_A < \infty]} + \frac{\sum_{z \in \partial A} \mathbb{P}_y[\tau_A^+ < \infty, S_{\tau_A^+} = z](G(y, x) - G(z, x))}{\mathbb{P}_x[\tau_A < \infty]}. \quad \text{(H.2)}$$

---

[2] I mean, only as in the first line of the first display there.

It is not difficult to check that, for distinct x, y, z (think about the situation when $x, y, z \in \mathbb{Z}^d$ are such that z is much closer to y than x is)

$$G(x, y) = G(x, z)(1 + O(\tfrac{\|y-z\|}{\|x-y\|})). \tag{H.3}$$

So, (3.14) implies that

$$\mathbb{P}_x[\tau_A < \infty] = \text{cap}(A)G(y, x)(1 + O(\tfrac{\text{diam}(A)}{\text{dist}(x,A)})), \tag{H.4}$$

which means that the first term in the right-hand side of (H.2) equals $\text{hm}_A(y)(1 + O(\tfrac{\text{diam}(A)}{\text{dist}(x,A)}))$.

To deal with the second term, denote $V = \partial B(y, 2\,\text{diam}(A) + 1)$; by e.g. (3.8), there exists $c > 0$ such that $\mathbb{P}_z[\tau_A = \infty] \geq c$ for all $z \in V$, and this permits us to obtain[3] that

$$\mathbb{P}_y[\tau_V < \tau_A^+] \leq c^{-1}\,\text{Es}_A(y). \tag{H.5}$$

We then use the optional stopping theorem with the martingale $M_n = G(y, x) - G(S_{n \wedge \tau_x}, x)$ and the stopping time $\tau = \tau_A^+ \wedge \tau_V$ to obtain that[4]

$$0 = \mathbb{E}_y M_0$$
$$= \mathbb{E}_y M_\tau$$

(note that $\mathbb{P}_y[\tau_A^+ < \tau_V] = \mathbb{P}_y[\tau_A^+ < \infty] - \mathbb{P}_y[\tau_A^+ > \tau_V, \tau_A^+ < \infty]$)

$$= \sum_{z \in \partial A} \mathbb{P}_y[\tau_A^+ < \infty, S_{\tau_A^+} = z](G(y, x) - G(z, x))$$
$$- \mathbb{P}_y[\tau_A^+ > \tau_V, \tau_A^+ < \infty](G(y, x) - \mathbb{E}_y(G(S_{\tau_V}) \mid \tau_A^+ > \tau_V, \tau_A^+ < \infty))$$
$$+ \mathbb{P}_y[\tau_V < \tau_A^+](G(y, x) - \mathbb{E}_y(G(S_{\tau_A^+}) \mid \tau_V < \tau_A^+)). \tag{H.6}$$

The first term in the right-hand side of (H.6) is what we need to estimate for (H.2), and, by (H.3) and (H.5), the second and third terms are both $O(\text{Es}_A(y)G(y, x)\tfrac{\text{diam}(A)}{\text{dist}(x,A)})$. Gathering the pieces, we obtain (3.79).

Exercise 3.26.

Start with almost (3.24):

$$\mathbb{P}_x[\tau_A < \infty, S_{\tau_A} = y] = G(y, x) - \sum_{z \in \partial A} \mathbb{P}_y[\tau_A^+ < \infty, S_{\tau_A^+} = z]G(z, x).$$

---

[3] Please elaborate.
[4] Of course, we also assume that x is outside $B(y, 2\,\text{diam}(A) + 1)$.

Then, consider the martingale $G(S_{n \wedge \tau_x}, x)$ and use the optional stopping theorem with the (possibly infinite[5]) stopping time $\tau_x \wedge \tau_A^+$ to obtain that

$$
\begin{aligned}
G(y, x) &= \mathbb{E}_y G(S_{\tau_x \wedge \tau_A^+}, x) \\
&= \sum_{z \in \partial A} \mathbb{P}_y[\tau_A^+ < \tau_x, S_{\tau_A^+} = z] G(z, x) + \mathbb{P}_y[\tau_x < \tau_A^+] G(0).
\end{aligned}
$$

So, using that

$$
\begin{aligned}
&\mathbb{P}_y[\tau_A^+ < \tau_x, S_{\tau_A^+} = z] \\
&= \mathbb{P}_y[\tau_A^+ < \infty, S_{\tau_A^+} = z] - \mathbb{P}_y[\tau_x < \tau_A^+ < \infty, S_{\tau_A^+} = z] \\
&= \mathbb{P}_y[\tau_A^+ < \infty, S_{\tau_A^+} = z] - \mathbb{P}_y[\tau_x < \tau_A^+] \mathbb{P}_x[\tau_A^+ < \infty, S_{\tau_A^+} = z],
\end{aligned}
$$

we obtain

$$
\begin{aligned}
&\mathbb{P}_x[\tau_A < \infty, S_{\tau_A} = y] \\
&= \mathbb{P}_y[\tau_x < \tau_A^+]\Big(G(0) - \sum_{z \in \partial A} \mathbb{P}_x[\tau_A^+ < \infty, S_{\tau_A^+} = z] G(z, x) \Big).
\end{aligned}
\tag{H.7}
$$

The second term in the preceding parentheses is clearly negative, and one can use (3.5) and (3.8) to obtain its magnitude and therefore show (3.80). To prove the second part, write

$$
\mathbb{P}_y[\tau_x < \infty, \tau_A^+ = \infty] = \mathbb{P}_y[\tau_x < \tau_A^+] \mathbb{P}_x[\tau_A = \infty],
$$

and then use (3.79) together with (3.4) and (3.8).

Exercise 3.28.

For (i) and (ii), see Figure H.2 (the left part for (i) and the right part for (ii)). Specifically, consider first a path starting from the origin and ending on the first hitting of the diagonal (respectively, vertical) line that separates the origin from $e_1 + e_2$ (respectively, x_1 from x_2). From the site y (respectively, z) where it ends, it is equally probable to go to 0 and x_1 (respectively, to x_1 and x_2). There are also paths that go from 0 to 0 (respectively, from 0 to x_1) which do not cross that line at all; this implies that the inequalities are strict.

As for part (iii), assume without restricting generality that x belongs to the first quadrant. Then, for an even x, the fact that $\mathbb{P}_0[S_{2n} = 0] > \mathbb{P}_0[S_{2n} = x]$ follows from (i) and (ii) by induction (there is a chain of sites that goes either from 0 or to $e_1 + e_2$ to x with "steps" $2e_1$ and $2e_2$). As for the case of

---

[5] We leave as an exercise checking that, in *this* case, one can still apply the optional stopping theorem; note that $G(S_\infty, x) = 0$ by transience.

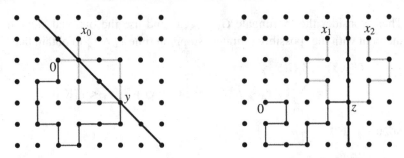

Figure H.2 For Exercise 3.28: on the left, $x_0 = e_1 + e_2$, on the right, $x_2 = x_1 + 2e_1$.

an odd x (where we need to prove that $\mathbb{P}_0[S_{2n} = 0] > \mathbb{P}_0[S_{2n+1} = x]$), use the fact that

$$\mathbb{P}_0[S_{2n+1} = x] = \frac{1}{4} \sum_{y \sim x} \mathbb{P}_0[S_{2n} = y]$$

and note that all y's in the preceding sum are even.

Exercise 3.30.

Hint: prove that τ_{Λ^C} has exponentially small tails using an argument of the sort "from any $y \in \Lambda$ the walker can go out of Λ in at most n_0 steps with probability at least $p_0 > 0$", with some explicit n_0 and p_0.

Exercise 3.33.

Indeed,

$$Y_{n \wedge \tau_{\Lambda^C}} = a(S_{n \wedge \tau_{\Lambda^C}} - y) - N_y^{(n \wedge \tau_{\Lambda^C} - 1)}$$

is still a martingale, so $a(x-y) = \mathbb{E}_x Y_{n \wedge \tau_{\Lambda^C}}$ for all n. Now, $\mathbb{E}_x a(S_{n \wedge \tau_{\Lambda^C}} - y) \to \mathbb{E}_x a(S_{\tau_{\Lambda^C}} - y)$ by the Dominated Convergence Theorem, and $\mathbb{E}_x N_y^{(n \wedge \tau_{\Lambda^C} - 1)} \to \mathbb{E}_x N_y^{(\tau_{\Lambda^C} - 1)}$ by the Monotone Convergence Theorem.

Exercise 3.34.

The idea is to use Theorem 3.13 with $\Lambda = B(R) \setminus \{0\}$, and then send R to infinity. Now, instead of just writing the arguments in a detailed way, let me show how one proves such statements *in practice*. Lemma 3.12 implies that $\mathbb{P}_x[\tau_0 \leq \tau_{B(R)^C}]$ is approximately $1 - \frac{a(x)}{a(R)}$, so, when $R \gg \|x\|$, Theorem 3.13 implies that

$$\mathbb{E}_x \eta_x \approx G_\Lambda(x, x) \approx a(x) \times \left(1 - \frac{a(x)}{a(R)}\right) + a(R) \times \frac{a(x)}{a(R)} \approx 2a(x).$$

For the second statement, write

$$\mathbb{E}_0 \eta_x = \frac{1}{4} \sum_{y \in \{\pm e_{1,2}\}} \mathbb{E}_y \eta_x$$

$$\approx \frac{1}{4} \sum_{y \in \{\pm e_{1,2}\}} G_\Lambda(y, x)$$

$$\approx \frac{1}{4} \sum_{y \in \{\pm e_{1,2}\}} \left(a(x) \times \left(1 - \frac{1}{a(R)}\right) + a(R) \times \frac{1}{a(R)} - a(x - y) \right)$$

$$\approx a(x) + 1 - \frac{1}{4} \sum_{z \sim x} a(z)$$

(since a is harmonic outside the origin)

$$= 1.$$

Inserting suitable O's and formally passing to the limits as $R \to \infty$ is *really* left as an exercise.

Exercise 3.35.

Indeed, conditioning on the location of the first entrance to B, we have

$$\mathbb{P}_x[S_{\tau_A} = y] = \sum_{z \in B} \mathbb{P}_x[S_{\tau_B} = z] \mathbb{P}_z[S_{\tau_A} = y].$$

Theorem 3.16 then implies that, as $x \to \infty$, the left-hand side converges to $\mathrm{hm}_A(y)$, and the right-hand side converges to

$$\sum_{z \in B} \mathrm{hm}_B(z) \mathbb{P}_z[S_{\tau_A} = y] = \mathbb{P}_{\mathrm{hm}_B}[S_{\tau_A} = y],$$

as required.

Exercise 3.36.

Hint for part (iii): when the walker is on the vertical axis, from that moment on it has equal chances of entering y or z first.

Exercise 3.41.

One method that would probably work is to first approximate the SRW with two-dimensional Brownian motion (possibly using a KMT-like strong approximation theorem [45, 56, 104]), and then find a suitable Möbius transform that sends the domain in Figure 3.8 to an annulus formed by *concentric* circumferences (and then use the conformal invariance of Brownian

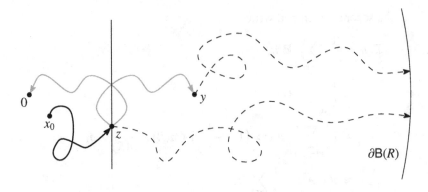

Figure H.3 On hitting a distant site y; note the symmetry of the grey trajectories and the fact that escaping from y or z to $\partial B(R)$ (dashed trajectories) happens approximately with the same probabilities.

trajectories). The author has to confess that he did not do any concrete calculations because he never needed such a result, but, nevertheless, it seems that this programme looks reasonably fine.

Exercise 4.6.

Hint: recall (3.49), the definition of q_A.

Exercise 4.10.

This is an easy consequence of the fact that the transition probabilities from x for the conditioned walk converge to those for the SRW as $x \to \infty$.

Exercise 4.11.

Let the $\widehat{S}$-walk start at x_0 (which is close to the origin), and consider a distant site y. Consider also the straight line such that 0 and y are symmetric with respect to it (see Figure H.3). As shown in the figure, it is likely that the walk first hits that line at some site z with $\|z\| \asymp \|y\|$; so, it would be enough to show that $\mathbb{P}_z[\hat{\tau}_y < \infty] \approx 1/2$. Then, take $R \gg \|y\|$; Lemma 3.12 then implies that $\mathbb{P}_y[\tau_{\partial B(R)} < \tau_0] \approx \frac{a(\|y\|)}{a(R)} \approx \mathbb{P}_z[\tau_{\partial B(R)} < \tau_0]$. On the other hand, by symmetry, $\frac{1}{2} = \mathbb{P}_z[\tau_y < \tau_0] \approx \mathbb{P}_z[\tau_y < \tau_0, \tau_{\partial B(R)} > \tau_{\{0,y\}}]$. This shows that

$$\mathbb{P}_z[\tau_y < \tau_{\partial B(R)} < \tau_0] \approx \frac{1}{2}\mathbb{P}_z[\tau_{\partial B(R)} < \tau_0]$$

and this is what we need indeed (recall Lemma 4.4).

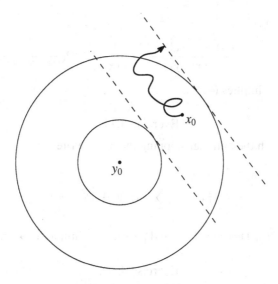

Figure H.4 Going out of the annulus.

This explanation is due to Yuval Peres.[6]

Exercise 4.12.

Hint: look at Figure H.4. Use the fact that the projection of the SRW on any fixed direction is a martingale, and then use the optional stopping theorem.

Alternatively: just notice that $\|S_{n \wedge \tau_{y_0}} - y_0\|$ is a submartingale.

Exercise 4.14.

Assume without restricting generality that $\|x\| \geq \|y\|$ (recall that $\hat{g}(x, y) = \hat{g}(y, x)$), and consider the following two cases.

Case 1: $\|y\| > \|x\|^{1/2}$. In this case, $a(x)$ and $a(y)$ are of the same order, and, since $\|x - y\| \leq 2\|x\|$, due to (3.36), $a(x - y) - a(x)$ is bounded above by a positive constant; therefore, the expression $a(x) + a(y) - a(x - y)$ will be of order $\ln \|x\|$. This implies that $\hat{g}(x, y)$ will be of order $\frac{1}{\ln \|x\|}$ indeed.

Case 2: $\|y\| \leq \|x\|^{1/2}$. Here, (3.38) implies that $a(x) - a(x - y) = O(\frac{\|x\|^{1/2}}{\|x\|}) =$

---

[6] On a conference, the author asked if anyone had a heuristic explanation of (4.21); Yuval was able to come up with it rather quickly.

$O(\|x\|^{-1/2})$, so

$$\hat{g}(x,y) = \frac{a(y) + O(\|x\|^{-1/2})}{a(x)a(y)} = \frac{1}{a(x)}\left(1 + O\left(\tfrac{1}{\|x\|^{1/2}\ln(1+\|y\|)}\right)\right),$$

and this again implies (4.86).

Exercise 4.17.

For $x \notin A$, with the optional stopping theorem, write

$$\hat{g}(x, y_0) = \mathbb{E}_x \hat{g}(\widehat{S}_{\hat{\tau}_A}, y_0)$$
$$= \mathbb{P}_x[\hat{\tau}_A < \infty] \sum_{z \in A} \hat{g}(z, y_0)\mathbb{P}_x[\widehat{S}_{\hat{\tau}_A} = z \mid \hat{\tau}_A < \infty],$$

then use (4.36), Theorem 4.16, and pass to the limit as $x \to \infty$.

Exercise 4.19.

Show that the process $Y_n = \widehat{G}(\widehat{S}_n, y) + \widehat{N}_y^{(n-1)}$ is a martingale and use the optional stopping theorem with the stopping time $\hat{\tau}_{A^C}$.

Exercise 4.21.

(i) Think about $A_n = B(r_n e_1, n)$, where $r_n \to \infty$ *very* quickly. Then, typically, if the conditioned walk ever visits $r_n e_1$, it will cover the whole A_n by "local recurrence" (recall that, when far away from the origin, the walk $\widehat{S}$ resembles very much the SRW by Lemma 4.5).

(ii) Think about

$$A_n = \{r_1 e_1, r_2 e_1, \ldots, r_n e_1\},$$

where (r_n) is as before,[7] and apply the reasoning similar to that used in the proof of Theorem 4.11.

Exercise 4.22.

This is open, I think (though it should not be *very* difficult).

Exercise 5.1.

Unfortunately, only with what one sees on Figure 5.6, it is not possible to find it. Think, for instance, that there may be a point just slightly above the top of the biggest triangle, which (I mean, the point) did not make it into the picture.

---

[7] The author thanks Hubert Lacoin for suggesting this example.

Exercise 5.2.

Due to Proposition 5.3, we can find a coupling $\mathbb{Q}$ between the Markov chain (X_i) and an i.i.d. collection (Y_i) (with law π), in such a way that for any $\lambda > 0$ and $t \geq 0$,

$$\mathbb{E}[\{Y_1, \ldots, Y_R\} \subset \{X_1, \ldots, X_t\}]$$

$$\geq \mathbb{P}_{\pi_0}\Big[\xi_0 \pi_0(x) + \sum_{j=1}^{t-1} \xi_j p(X_j, x) \geq \lambda \pi(x), \text{ for all } x \in \Sigma\Big], \quad (\text{H.8})$$

where ξ_i are i.i.d. Exp(1) random variables, independent of R, a Poisson(λ)-distributed random variable. Then obtain from a simple calculation the fact that $\mathbb{P}[\{Y_1, \ldots, Y_R\} \cap A = \emptyset] = e^{-\lambda\pi(A)}$.

Exercise 5.3.

If you find one, please, let me know.

Exercise 5.6.

To the second one. See also [53] for a more complete discussion of this.

Exercise 5.10.

Think how one can generate the lines in the order corresponding

- to the distances from the origin to the lines;
- to the distances from the origin points of intersection with the horizontal axis.

Exercise 5.15.

Answer: 3/2. It is the last problem of the famous Mathematical Trivium [4] of Vladimir Arnold. Clearly, the problem reduces to finding the expected area of a projection of a square (note that a.s. only three faces of the cube contribute to the projection), and then one can calculate the answer doing a bit of integration. There is another way to solve it, however, that does not require any computations at all, and works for any three-dimensional convex set, not only for the cube. One may reason in the following way:

- imagine the surface of a convex set to be composed of many small plaquettes, and use the linearity of expectation to argue that the expected area of the projection *equals* the surface area of the set times a constant (that is, it does not depend on the surface's shape itself!);

- to obtain this constant, consider a certain special convex set whose projections are the same in all directions (finding such a set is left as an exercise).

Exercise 6.1.

With (6.5) and (6.6), it is straightforward to obtain that $\varepsilon = \frac{G(x-y)}{G(0)+G(x-y)}$; it is of order $\|x - y\|^{-(d-2)}$ as $\|x - y\| \to \infty$.

Exercise 6.2.

See the proof of (6.2) and (6.4) in [79].

Exercise 6.8.

Hint: use the fact that $\mathcal{V}^{\alpha+\varepsilon}$ has the same law as $\mathcal{V}^{\alpha} \cap \widetilde{\mathcal{V}}^{\varepsilon}$, where $\widetilde{\mathcal{V}}^{\varepsilon}$ is an independent (of the whole family $(\mathcal{V}^s, s > 0)$) copy of $\mathcal{V}^{\varepsilon}$.

Exercise 6.12.

See [15].

References

[1] Abe, Yoshihiro. 2017. Second order term of cover time for planar simple random walk. *arXiv preprint arXiv:1709.08151*. (Cited on pages 160 and 167.)

[2] Abe, Yoshihiro, and Biskup, Marek. 2019. Exceptional points of two-dimensional random walks at multiples of the cover time. *arXiv preprint arXiv:1903.04045*. (Cited on page 168.)

[3] Alves, Caio, and Popov, Serguei. 2018. Conditional decoupling of random interlacements. *ALEA Lat. Am. J. Probab. Math. Stat.*, **15**(2), 1027–1063. (Cited on page 159.)

[4] Arnold, V. I. 1992. Trivium mathématique. *Gaz. Math.*, 87–96. (Cited on page 199.)

[5] Asmussen, S. 2003. *Applied probability and queues*. Second edn. Applications of Mathematics (New York), vol. 51. Springer-Verlag, New York. (Cited on page 17.)

[6] Belius, David. 2012. Cover levels and random interlacements. *Ann. Appl. Probab.*, **22**(2), 522–540. (Cited on page 149.)

[7] Belius, David. 2013. Gumbel fluctuations for cover times in the discrete torus. *Probab. Theory Related Fields*, **157**(3–4), 635–689. (Cited on page 160.)

[8] Belius, David, and Kistler, Nicola. 2017. The subleading order of two dimensional cover times. *Probab. Theory Related Fields*, **167**(1–2), 461–552. (Cited on pages 160 and 167.)

[9] Benjamini, Itai, and Sznitman, Alain-Sol. 2008. Giant component and vacant set for random walk on a discrete torus. *J. Eur. Math. Soc. (JEMS)*, **10**(1), 133–172. (Cited on page 146.)

[10] Berger, Noam, Gantert, Nina, and Peres, Yuval. 2003. The speed of biased random walk on percolation clusters. *Probab. Theory Related Fields*, **126**(2), 221–242. (Cited on page 147.)

[11] Bollobás, Béla, and Riordan, Oliver. 2006. *Percolation*. Cambridge University Press, New York. (Cited on page 146.)

[12] Boucheron, S., Lugosi, G., and Massart, P. 2013. *Concentration inequalities : a nonasymptotic theory of independence*. Oxford University Press, Oxford. (Cited on page 125.)

[13] Brémaud, Pierre. 1999. *Markov chains*. Texts in Applied Mathematics, vol. 31. Springer-Verlag, New York. (Cited on page 17.)

[14] Bucy, R. S. 1965. Recurrent sets. *Ann. Math. Statist.*, **36**, 535–545. (Cited on page 190.)

[15] Camargo, Darcy, and Popov, Serguei. 2018. A one-dimensional version of the random interlacements. *Stochastic Processes and Their Applications*, **128**(8), 2750–2778. (Cited on page 200.)

[16] Černý, J., and Popov, S. 2012. On the internal distance in the interlacement set. *Electron. J. Probab.*, **17**, paper no. 29, 25 pp. (Cited on pages 146 and 147.)

[17] Černý, Jiří, and Teixeira, Augusto. 2012. *From random walk trajectories to random interlacements*. Ensaios Matemáticos [Mathematical Surveys], vol. 23. Sociedade Brasileira de Matemática, Rio de Janeiro. (Cited on page 143.)

[18] Černý, Jiří, and Teixeira, Augusto. 2016. Random walks on torus and random interlacements: macroscopic coupling and phase transition. *Ann. Appl. Probab.*, **26**(5), 2883–2914. (Cited on page 146.)

[19] Chung, Kai Lai, and Walsh, John B. 2005. *Markov processes, Brownian motion, and time symmetry*. Vol. 249. Springer, New York. (Cited on page 75.)

[20] Comets, F., Popov, S., Schütz, G. M., and Vachkovskaia, M. 2009. Billiards in a general domain with random reflections. *Arch. Ration. Mech. Anal.*, **191**(3), 497–537. (Cited on page 140.)

[21] Comets, F., Popov, S., Schütz, G. M., and Vachkovskaia, M. 2010a. Knudsen gas in a finite random tube: transport diffusion and first passage properties. *J. Stat. Phys.*, **140**(5), 948–984. (Cited on page 140.)

[22] Comets, F., Popov, S., Schütz, G. M., and Vachkovskaia, M. 2010b. Quenched invariance principle for the Knudsen stochastic billiard in a random tube. *Ann. Probab.*, **38**(3), 1019–1061. (Cited on page 140.)

[23] Comets, Francis, and Popov, Serguei. 2017. The vacant set of two-dimensional critical random interlacement is infinite. *Ann. Probab.*, **45**(6B), 4752–4785. (Cited on pages 125, 127, and 167.)

[24] Comets, Francis, Menshikov, Mikhail, and Popov, Serguei. 1998. Lyapunov functions for random walks and strings in random environment. *Ann. Probab.*, **26**(4), 1433–1445. (Cited on page 27.)

[25] Comets, Francis, Gallesco, Christophe, Popov, Serguei, and Vachkovskaia, Marina. 2013. On large deviations for the cover time of two-dimensional torus. *Electron. J. Probab.*, **18**, paper no. 96, 18 pp. (Cited on pages 117 and 169.)

[26] Comets, Francis, Popov, Serguei, and Vachkovskaia, Marina. 2016. Two-dimensional random interlacements and late points for random walks. *Comm. Math. Phys.*, **343**(1), 129–164. (Cited on pages 10, 80, 117, 173, and 175.)

[27] Cook, Scott, and Feres, Renato. 2012. Random billiards with wall temperature and associated Markov chains. *Nonlinearity*, **25**(9), 2503–2541. (Cited on page 140.)

[28] Daley, D. J., and Vere-Jones, D. 2008. *An introduction to the theory of point processes*. Vol. II. Springer, New York. (Cited on page 146.)

[29] de Bernardini, Diego, Gallesco, Christophe, and Popov, Serguei. 2018a. An improved decoupling inequality for random interlacements. *arXiv preprint arXiv:1809.05594*. (Cited on pages 149 and 159.)

[30] de Bernardini, Diego F., Gallesco, Christophe, and Popov, Serguei. 2018b. On uniform closeness of local times of Markov chains and i.i.d. sequences. *Stochastic Process. Appl.*, **128**(10), 3221–3252. (Cited on pages 125, 127, and 131.)

[31] de Bernardini, Diego F., Gallesco, Christophe, and Popov, Serguei. 2019. Corrigendum to "On uniform closeness of local times of Markov chains and i.i.d. sequences" [Stochastic Process. Appl. 128 (10) 3221–3252]. *Stochastic Processes and Their Applications*, **129**(7), 2606–2609. (Cited on pages 127 and 131.)

[32] Dembo, Amir, and Sznitman, Alain-Sol. 2006. On the disconnection of a discrete cylinder by a random walk. *Probab. Theory Related Fields*, **136**(2), 321–340. (Cited on page 125.)

[33] Dembo, Amir, Peres, Yuval, Rosen, Jay, and Zeitouni, Ofer. 2004. Cover times for Brownian motion and random walks in two dimensions. *Ann. of Math. (2)*, **160**(2), 433–464. (Cited on page 160.)

[34] Dembo, Amir, Peres, Yuval, Rosen, Jay, and Zeitouni, Ofer. 2006. Late points for random walks in two dimensions. *Ann. Probab.*, **34**(1), 219–263. (Cited on pages 64, 160, and 172.)

[35] Di Crescenzo, A., Macci, C., Martinucci, B., and Spina, S. 2017. Random walks on graphene: generating functions, state probabilities and asymptotic behavior. *arXiv:1610.09310*. (Cited on page 185.)

[36] Ding, Jian, Lee, James R., and Peres, Yuval. 2012. Cover times, blanket times, and majorizing measures. *Ann. of Math. (2)*, **175**(3), 1409–1471. (Cited on page 125.)

[37] Doob, J. L. 1957. Conditional Brownian motion and the boundary limits of harmonic functions. *Bull. Soc. Math. France*, **85**, 431–458. (Cited on page 75.)

[38] Doob, J. L. 1984. *Classical potential theory and its probabilistic counterpart*. Grundlehren der Mathematischen Wissenschaften [Fundamental Principles of Mathematical Sciences], vol. 262. Springer-Verlag, New York. (Cited on pages 33 and 75.)

[39] Doyle, P. G., and Snell, J. L. 1984. *Random walks and electric networks*. Carus Mathematical Monographs, vol. 22. Mathematical Association of America, Washington, DC. (Cited on page 14.)

[40] Drewitz, Alexander, and Erhard, Dirk. 2016. Transience of the vacant set for near-critical random interlacements in high dimensions. *Ann. Inst. Henri Poincaré Probab. Stat.*, **52**(1), 84–101. (Cited on page 147.)

[41] Drewitz, Alexander, Ráth, Balázs, and Sapozhnikov, Artëm. 2014a. *An introduction to random interlacements*. Springer Briefs in Mathematics. Springer, Cham. (Cited on pages 8 and 143.)

[42] Drewitz, Alexander, Ráth, Balázs, and Sapozhnikov, Artëm. 2014b. Local percolative properties of the vacant set of random interlacements with small intensity. *Ann. Inst. Henri Poincaré Probab. Stat.*, **50**(4), 1165–1197. (Cited on page 147.)

[43] Drewitz, Alexander, Ráth, Balázs, and Sapozhnikov, Artëm. 2014c. On chemical distances and shape theorems in percolation models with long-range correlations. *J. Math. Phys.*, **55**(8), 083307, 30. (Cited on page 147.)

[44] Durrett, R. 2010. *Probability: theory and examples*. Fourth edn. Cambridge Series in Statistical and Probabilistic Mathematics. Cambridge University Press, Cambridge. (Cited on pages 8, 4, and 5.)

[45] Einmahl, Uwe. 1989. Extensions of results of Komlós, Major, and Tusnády to the multivariate case. *Journal of Multivariate Analysis*, **28**(1), 20–68. (Cited on page 195.)

[46] Fayolle, G., Malyshev, V. A., and Menshikov, M. V. 1995. *Topics in the constructive theory of countable Markov chains.* Cambridge University Press, Cambridge. (Cited on pages 17, 19, and 30.)

[47] Fribergh, Alexander, and Popov, Serguei. 2018. Biased random walks on the interlacement set. *Ann. Inst. Henri Poincaré Probab. Stat.*, **54**(3), 1341–1358. (Cited on page 147.)

[48] Gantert, Nina, Popov, Serguei, and Vachkovskaia, Marina. 2019. On the range of a two-dimensional conditioned simple random walk. *Annales Henri Lebesgue*, **2**, 349–368. (Cited on page 101.)

[49] Garcia, Nancy L. 1995. Birth and death processes as projections of higher-dimensional Poisson processes. *Adv. in Appl. Probab.*, **27**(4), 911–930. (Cited on page 117.)

[50] Garcia, Nancy L., and Kurtz, Thomas G. 2008. Spatial point processes and the projection method. Pages 271–298 of: *In and out of equilibrium. 2.* Progr. Probab., vol. 60. Birkhäuser, Basel. (Cited on page 117.)

[51] Grimmett, Geoffrey. 1999. *Percolation.* Second edn. Grundlehren der Mathematischen Wissenschaften [Fundamental Principles of Mathematical Sciences], vol. 321. Springer-Verlag, Berlin. (Cited on page 146.)

[52] Hu, Yueyun, and Shi, Zhan. 2015. The most visited sites of biased random walks on trees. *Electron. J. Probab.*, **20**, paper no. 62, 14. (Cited on page 125.)

[53] Jaynes, E. T. 1973. The well-posed problem. *Found. Phys.*, **3**, 477–492. (Cited on page 199.)

[54] Kendall, M. G., and Moran, P. A. P. 1963. *Geometrical probability.* Griffin's Statistical Monographs & Courses, No. 10. Hafner Publishing Co., New York. (Cited on page 136.)

[55] Kochen, Simon, and Stone, Charles. 1964. A note on the Borel–Cantelli lemma. *Illinois J. Math.*, **8**, 248–251. (Cited on page 110.)

[56] Komlós, János, Major, P'eter, and Tusnády, G. 1975. An approximation of partial sums of independent RV'-s, and the sample DF. I. *Zeitschrift für Wahrscheinlichkeitstheorie und Verwandte Gebiete*, **32**, 111–131. (Cited on page 195.)

[57] Lacoin, Hubert, and Tykesson, Johan. 2013. On the easiest way to connect k points in the random interlacements process. *ALEA Lat. Am. J. Probab. Math. Stat.*, **10**(1), 505–524. (Cited on page 146.)

[58] Lagarias, J. C., and Weiss, A. 1992. The $3x + 1$ problem: two stochastic models. *Ann. Appl. Probab.*, **2**(1), 229–261. (Cited on page 31.)

[59] Lamperti, J. 1960. Criteria for the recurrence or transience of stochastic process. I. *J. Math. Anal. Appl.*, **1**, 314–330. (Cited on page 32.)

[60] Lamperti, J. 1962. A new class of probability limit theorems. *J. Math. Mech.*, **11**, 749–772. (Cited on page 32.)

[61] Lamperti, J. 1963. Criteria for stochastic processes. II. Passage-time moments. *J. Math. Anal. Appl.*, **7**, 127–145. (Cited on page 32.)

[62] LaSalle, Joseph, and Lefschetz, Solomon. 1961. *Stability by Liapunov's direct method, with applications.* Mathematics in Science and Engineering, Vol. 4. Academic Press, New York–London. (Cited on page 18.)

[63] Lawler, G. F., and Limic, V. 2010. *Random walk: a modern introduction.* Cambridge Studies in Advanced Mathematics, vol. 123. Cambridge University Press, Cambridge. (Cited on pages 8, 9, 33, 36, 46, 52, 65, 69, and 188.)

[64] Levin, David A, and Peres, Yuval. 2017. *Markov chains and mixing times*. Vol. 107. American Mathematical Society, Providence. (Cited on pages 14 and 75.)

[65] Liggett, T. M. 1985. *Interacting particle systems*. Grundlehren der Mathematischen Wissenschaften [Fundamental Principles of Mathematical Sciences], vol. 276. Springer-Verlag, New York. (Cited on page 29.)

[66] Lyons, Russell, and Peres, Yuval. 2016. *Probability on trees and networks*. Cambridge Series in Statistical and Probabilistic Mathematics, vol. 42. Cambridge University Press, New York. (Cited on pages 14 and 29.)

[67] Meester, Ronald, and Roy, Rahul. 1996. *Continuum percolation*. Cambridge Tracts in Mathematics, vol. 119. Cambridge University Press, Cambridge. (Cited on page 132.)

[68] Menshikov, M., Popov, S., Ramírez, A., and Vachkovskaia, M. 2012. On a general many-dimensional excited random walk. *Ann. Probab.*, **40**(5), 2106–2130. (Cited on page 27.)

[69] Menshikov, M. V. 1986. Coincidence of critical points in percolation problems. *Dokl. Akad. Nauk SSSR*, **288**(6), 1308–1311. (Cited on page 147.)

[70] Menshikov, M. V., and Popov, S. Yu. 1995. Exact power estimates for countable Markov chains. *Markov Process. Related Fields*, **1**(1), 57–78. (Cited on page 27.)

[71] Menshikov, Mikhail, Popov, Serguei, and Wade, Andrew. 2017. *Non-homogeneous random walks – Lyapunov function methods for near-critical stochastic systems*. Cambridge University Press, Cambridge. (Cited on pages 8, 7, 17, 21, 27, 30, 186, and 187.)

[72] Meyn, S., and Tweedie, R. L. 2009. *Markov chains and stochastic stability*. Second edn. Cambridge University Press, Cambridge. With a prologue by Peter W. Glynn. (Cited on page 17.)

[73] Miller, Jason, and Sousi, Perla. 2017. Uniformity of the late points of random walk on $\mathbb{Z}_n^d$ for $d \geq 3$. *Probab. Theory Related Fields*, **167**(3–4), 1001–1056. (Cited on page 160.)

[74] Peres, Y., Popov, S., and Sousi, P. 2013. On recurrence and transience of self-interacting random walks. *Bull. Braz. Math. Soc.*, **44**(4), 1–27. (Cited on page 21.)

[75] Peres, Yuval. 2009. The unreasonable effectiveness of martingales. Pages 997–1000 of: *Proceedings of the Twentieth Annual ACM-SIAM Symposium on Discrete Algorithms*. SIAM, Philadelphia. (Cited on page 4.)

[76] Pólya, Georg. 1921. Über eine Aufgabe der Wahrscheinlichkeitsrechnung betreffend die Irrfahrt im Straßennetz. *Math. Ann.*, **84**(1–2), 149–160. (Cited on pages 1 and 27.)

[77] Popov, Serguei. 2019. Conditioned two-dimensional simple random walk: Green's function and harmonic measure. *arXiv preprint arXiv:1907.12682*. (Cited on pages 81 and 92.)

[78] Popov, Serguei, and Ráth, Balázs. 2015. On decoupling inequalities and percolation of excursion sets of the Gaussian free field. *J. Stat. Phys.*, **159**(2), 312–320. (Cited on page 149.)

[79] Popov, Serguei, and Teixeira, Augusto. 2015. Soft local times and decoupling of random interlacements. *J. Eur. Math. Soc. (JEMS)*, **17**(10), 2545–2593. (Cited on pages 117, 119, 121, 124, 147, 150, 151, 156, 157, 158, and 200.)

[80] Popov, Serguei, Rolla, Leonardo T., and Ungaretti, Daniel. 2020. Transience of conditioned walks on the plane: encounters and speed of escape. *Electron. J. Probab.*, **25**, paper no. 52, 23 pp. (Cited on page 185.)

[81] Procaccia, Eviatar B., and Tykesson, Johan. 2011. Geometry of the random interlacement. *Electron. Commun. Probab.*, **16**, 528–544. (Cited on page 146.)

[82] Procaccia, Eviatar B., Rosenthal, Ron, and Sapozhnikov, Artëm. 2016. Quenched invariance principle for simple random walk on clusters in correlated percolation models. *Probab. Theory Related Fields*, **166**(3–4), 619–657. (Cited on page 147.)

[83] Ráth, Balázs, and Sapozhnikov, Artëm. 2012. Connectivity properties of random interlacement and intersection of random walks. *ALEA Lat. Am. J. Probab. Math. Stat.*, **9**, 67–83. (Cited on page 146.)

[84] Ráth, Balázs, and Sapozhnikov, Artëm. 2013. The effect of small quenched noise on connectivity properties of random interlacements. *Electron. J. Probab.*, **18**, paper no. 4, 20 pp. (Cited on page 147.)

[85] Resnick, Sidney I. 2008. *Extreme values, regular variation and point processes.* Springer Series in Operations Research and Financial Engineering. Springer, New York. Reprint of the 1987 original. (Cited on page 117.)

[86] Revuz, D. 1984. *Markov chains.* Second edn. North-Holland Mathematical Library, vol. 11. Amsterdam: North-Holland Publishing Co. (Cited on pages 33 and 111.)

[87] Rodriguez, Pierre-François. 2019. On pinned fields, interlacements, and random walk on $(\mathbb{Z}/N\mathbb{Z})^2$. *Probab. Theory Related Fields*, **173**(3), 1265–1299. (Cited on page 168.)

[88] Romeijn, H. E. 1998. A general framework for approximate sampling with an application to generating points on the boundary of bounded convex regions. *Statist. Neerlandica*, **52**(1), 42–59. (Cited on page 139.)

[89] Sapozhnikov, Artem. 2017. Random walks on infinite percolation clusters in models with long-range correlations. *Ann. Probab.*, **45**(3), 1842–1898. (Cited on page 147.)

[90] Sidoravicius, Vladas, and Sznitman, Alain-Sol. 2009. Percolation for the vacant set of random interlacements. *Comm. Pure Appl. Math.*, **62**(6), 831–858. (Cited on page 147.)

[91] Spitzer, F. 1976. *Principles of random walk.* Second edn. Springer-Verlag, New York. Graduate Texts in Mathematics, Vol. 34. (Cited on pages 8, 53, and 69.)

[92] Srikant, R., and Ying, Lei. 2014. *Communication Networks: An Optimization, Control and Stochastic Networks Perspective.* Cambridge University Press, New York. (Cited on page 17.)

[93] Sznitman, Alain-Sol. 2010. Vacant set of random interlacements and percolation. *Ann. of Math. (2)*, **171**(3), 2039–2087. (Cited on pages 10, 125, 143, 145, and 149.)

[94] Sznitman, Alain-Sol. 2012a. On $(\mathbb{Z}/N\mathbb{Z})^2$-occupation times, the Gaussian free field, and random interlacements. *Bull. Inst. Math. Acad. Sin. (N.S.)*, **7**(4), 565–602. (Cited on page 149.)

[95] Sznitman, Alain-Sol. 2012b. Random interlacements and the Gaussian free field. *Ann. Probab.*, **40**(6), 2400–2438. (Cited on page 149.)

[96] Sznitman, Alain-Sol. 2012c. *Topics in occupation times and Gaussian free fields.* Zurich Lectures in Advanced Mathematics. European Mathematical Society (EMS), Zürich. (Cited on page 143.)

[97] Sznitman, Alain-Sol. 2017. Disconnection, random walks, and random interlacements. *Probab. Theory Related Fields*, **167**(1–2), 1–44. (Cited on page 10.)

[98] Teixeira, A. 2009a. Interlacement percolation on transient weighted graphs. *Electron. J. Probab.*, **14**, paper no. 54, 1604–1628. (Cited on pages 150, 162, and 163.)

[99] Teixeira, Augusto. 2009b. On the uniqueness of the infinite cluster of the vacant set of random interlacements. *Ann. Appl. Probab.*, **19**(1), 454–466. (Cited on page 147.)

[100] Teixeira, Augusto. 2011. On the size of a finite vacant cluster of random interlacements with small intensity. *Probab. Theory Related Fields*, **150**(3–4), 529–574. (Cited on page 147.)

[101] Tóth, Bálint. 2001. No more than three favorite sites for simple random walk. *Ann. Probab.*, **29**(1), 484–503. (Cited on page 125.)

[102] Tsirelson, Boris. 2006. Brownian local minima, random dense countable sets and random equivalence classes. *Electron. J. Probab.*, **11**, paper no. 7, 162–198. (Cited on page 117.)

[103] Woess, Wolfgang. 2009. *Denumerable Markov chains.* EMS Textbooks in Mathematics. European Mathematical Society (EMS), Zürich. (Cited on page 75.)

[104] Zaitsev, Andrei Yu. 1998. Multidimensional version of the results of Komlos, Major and Tusnady for vectors with finite exponential moments. *ESAIM: PS*, **2**, 41–108. (Cited on page 195.)

Index

adapted process, 3
Bessel process, 77
binomial process, 131
capacity
  for SRW in $d \geq 3$, 37
  for the conditioned SRW, 88
  for two-dimensional SRW, 62
  of d-dimensional ball, 40
  of a disk, 63
  of one- and two-point sets, 39, 63
  of three-point sets, 70
  relationship of two-dimensional
    capacities, 88
Chapman–Kolmogorov equation, 5
conductance, 13
cosine reflection law, 135, 139
coupling, 34, 46–49, 104, 120, 121, 124,
  125, 131, 151, 152, 179, 199
cover time, 125, 160

decoupling, 11, 27, 69, 140, 151, 152
Doob's h-transform, 75
drift (with respect to a Lyapunov
  function), 21

effective resistance, 14
electrical network, 13
empirical processes, 125
escape probability
  for SRW in $d \geq 3$, 37
  for the conditioned SRW, 85
excursions, 80, 101, 103, 104, 107, 151,
  153
filtration, 3
Foster's criterion, 30
Foster–Lyapunov theorem, 30

Gambler's Ruin Problem, 73
Green's function
  asymptotic behaviour, 36, 56, 65

exact expression, 65
for SRW in $d \geq 3$, 35
for the conditioned SRW, 81, 87, 112
relation to the potential kernel, 55
restricted on a set, 54, 55, 113

Hölder continuity, 128
harmonic function, 33
harmonic measure
  as entrance measure from infinity, 43,
    58, 60, 97
  consistency, 67, 69, 142, 143
  for SRW in $d \geq 3$, 42
  for the conditioned SRW, 92, 97
  for two-dimensional SRW, 57, 58, 60
harmonic series, 9, 11, 16, 134
hitting time, 3, 6

irreducibility, 5

Kochen–Stone theorem, 110

Lamperti problem, 32
last-visit decomposition, 38, 86
local time, 123, 125, 127, 129
Lyapunov function, 17, 18, 21
Lyapunov function criterion
  for positive recurrence, 30
  for recurrence, 17
  for transience, 20

magic lemma, 117
Markov chain
  aperiodic, 6
  criterion for positive recurrence, 30
  criterion for recurrence, 17
  criterion for reversibility, 12
  criterion for transience, 20
  discrete-time, 5
  invariant measure, 6
  irreducible, 5
  null recurrent, 6

Printed in the United States
by Baker & Taylor Publisher Services

Printed in the United States
by Baker & Taylor Publisher Services